ADVANCE PRAISE *for* URGENT MESSAGE *from* MOTHER

"Mom says: clean your room. What do Gloria Steinem and Archbishop Desmond Tutu have in common? They're both listening to an urgent message from Mother Earth, as are a host of others who are singing the praises of Bolen's latest book"

—*Utne*

"Bolen provides the tools for women to lead the world away from the perils of patriarchy and toward life-sustaining change."

—*Ode*

"The Mother has sent an urgent message. Are you ready to answer the call?"

—*Spirituality & Health*

"This is the most inspiring and optimistic book I've read in years. It tells how women working together can bring us peace and save the planet. Jean Shinoda Bolen invites us all to join the next, most powerful wave of the women's movement. Count me in!"

—Isabel Allende, author of *Zorro, The House of the Spirits*, and *Paula*

"Jean Shinoda Bolen shows us how the cult of masculinity is endangering us all. Women and men are equally human and fallible but at least women don't have our masculinity to prove—and that alone may make us the main saviors of this fragile Spaceship Earth."

—**Gloria Steinem**, cofounder of *Ms.* magazine and the Ms. Foundation for Women

"Jean Shinoda Bolen's *Urgent Message from Mother* is a book whose time has come. Our earth home and all forms of life in it are at grave risk. We men have had our turn and made a proper mess of things. We need women to save us. I pray that many will read Bolen's work and be inspired then to act appropriately. Time is running out."

—**Archbishop Desmond Tutu**, recipient of the Nobel Peace Prize, Chair of the Truth and Reconciliation Commission of South Africa

"Jean Shinoda Bolen's Urgent Message from Mother aims at a day that is coming but has not arrived. A day when religions will not be exclusively patriarchal, when makers of war will first ask questions which mothers ask, when societies will measure success factoring in the health of our planet as of the first order! This book emboldens readers by referring to ancient wisdom that has been muted and by aspiring to a way of living in the future that includes the feminine genius. Obviously the critical matter of garnering and wielding responsible power must be addressed. Bolen takes beginning steps in that direction; more will follow, no doubt. But *Urgent Message from Mother* moves our attention upward to a time of needed fundamental change in the way people of this world think and act. It is a foundational book and a strong one."

—**The Right Reverend William E. Swing**, Episcopal Bishop of California and President of the United Religions Initiative

"Jean Shinoda Bolen's potent *Urgent Message from Mother* has the empowerment to be the right catalyst for the right time. This book will most certainly motivate women to gather together to be spiritually uplifted and to bring to the world the love and peace that are so sadly lacking in the world we see today."

—**Gerald G. Jampolsky**, M.D., founder, Center for Attitudinal Healing

"There has been a major change in the world since the end of World War II. While many problems still affect humanity, the Earth, our Mother, has become priority number one. Jean Shinoda Bolen's *Urgent Message from Mother* comes as a true blessing. May it receive a wide audience and have a profound impact."

—**Dr. Robert Muller**, former Assistant Secretary General of the United Nations and winner of the 1989 UNESCO Peace Education Prize

"An inspiring call to action, powered by real examples of how women can save the world."

—**Gloria Feldt**, author of *The War on Choice: The Right-Wing Attack on Women's Rights and How to Fight Back* and former president, Planned Parenthood Federation of America

"Jean takes feminist spirituality to the next level—collective activism. *Urgent Message from Mother* is a timely and critical call for a new wave of feminism that will inspire and galvanize women and girls everywhere via interlaced circles. *Urgent Message* tells us that our circles are the mechanism for positive change and healing in a world of fear and fundamentalism. Since women were the first democracy builders, gathering in circles since the beginning of time, we understand the invisible power of the conscious circle as the polar opposite of hierarchy. I applaud Jean's call to action!"

—**Marilyn Fowler**, President of the Women's Intercultural Network and State Coordinator of the California Women's Agenda

JEAN SHINODA BOLEN, M.D.

URGENT MESSAGE FROM MOTHER

Gather the Women, Save the World

First published in 2005 by Conari Press,
an imprint of Red Wheel/Weiser, LLC
With offices at:
500 Third Street, Suite 230
San Francisco, CA 94107
www.redwheelweiser.com

Library of Congress Cataloging-in-Publication Data

Bolen, Jean Shinoda.
Urgent message from mother : gather the women, save the world / Jean Shinoda Bolen.
p. cm.
Includes bibliographical references and index.
ISBN 978-1-57324-353-7 (alk. paper)
1. Women—Psychology. 2. Women—Social networks. 3. Women political activists. I. Title.
HQ1206.B555 2005
305.4—dc22
2005011828

Typeset in Minion by Kathleen Wilson Fivel

Printed in the United States of America
MV
10 9 8 7 6 5 4 3 2 1

The paper used in this publication meets the minimum requirements of the American National Standard for Information Sciences—Permanence of Paper for Printed Library Materials Z39.48-1992 (R1997).

DEDICATION

For my daughter Melody Jean Bolen
and
my son Andre Joseph Bolen
(February 16, 1972–June 4, 2001)

CONTENTS

PREFACE TO THE PAPERBACK EDITION

HEEDING THE URGENT MESSAGE FROM MOTHER

In the short period between the initial 2005 publication of *Urgent Message from Mother: Gather the Women, Save the World* and this paperback edition, there have been many hopeful signs and good people that encourage me to continue to be a message carrier. I find that those who are heeding a Message from Mother are motivated by love for whatever it is that needs to be nourished, protected, or saved. It's advocacy, caretaking, or outrage at indifference or abuse of power. These people are putting their efforts into saving the life and well-being of a particular vulnerable living thing or thing of beauty or into saving the world.

My optimism was given a boost when Global Warming went mainstream in the time between the hardcover and softcover editions. A tipping point was reached—and as if, overnight—global warming was an idea whose time had come. An infectious idea causes widespread changes in perception and behavior when a critical number of people adopt it. Green is now commercial. Hybrid cars are hot. Yet long before a movement surfaces or a resisted idea becomes accepted, innumerable individuals had to commit themselves to do something they felt called to do. Think of Al Gore before the documentary and the Nobel Prize, taking his power-point slide show around, a figure of ridicule to the powerbrokers. He was like all the many others who believe in the importance of what they are doing in the absence of evidence that it is making a difference. It is necessary for many, many people to do grassroots grunt work until a tipping point is reached. This is the hundredth monkey or millionth circle principle, and how what was once not done, not believed, or unthinkable becomes what most people accept as reality.

In this same period between the hardcover edition and this one, the tenth anniversary edition of another one of my books, *Close to the Bone: Life-Threatening Illness as a Soul Journey*, was published. I see strong parallels between how a person responds to a cancer diagnosis and what people do in response to dire news about what is happening in their world or the world. Some patients give up and expect the worst prognosis as inevitable. Most accept that those in authority are the experts and leave it up to them. Then there is a significant minority who are exceptional. They act on the belief that they can make a difference in the outcome and seek information, second and third opinions, and look into alternatives. They may do inner work, find support groups and make major changes in how they live. Exceptional patients do whatever they believe will help them get well.

People who become advocates for good causes are exceptional in similar ways. Most non-governmental organizations are formed by such people. The proliferation of NGOs in the past decade is amazing: they doubled in the United States to over a million, grew from none to speak of to several hundred thousand in China and Russia. India now has a half million NGOs.

Bishop Desmond Tutu wrote, "Our earth home and all forms of life in it are at grave risk. We men have had our turn and made a proper mess of things. We need women to save us." Some recognition of this may be reflected in the increasing number of countries that have women heads of state: there were twenty-one in 2007.

Women are the empathic gender, which doesn't mean every woman, but it does mean that women collectively have this as a distinguishing quality. Empathy is the understanding that what hurts me would feel the same way to you. It is the foundation upon which moral decisions are made. In chapter 4, "Mother Needs *You!,*" I describe characteristic qualities that women have (as do exceptional men). The point is that these talents are needed now.

In 2006, research on the female brain got widespread media attention and was the subject of a best-selling book. These studies found that there are four parts of the brain that are usually larger and more active in women than men. These parts have to do with self control, patience, intuition, empathy, ability to weigh options, worry, and emotional memory.

The message contained in this book's subtitle, "Gather the Women, Save the World," is about the need for women to participate in peacemaking at every level and has to do with feminism, spiritual activism, mother power, women in circles with a sacred center, and three to the nineteenth power. If three women heed the message and if they tell three other women, there would be nine. If each of them then spread the word to three others, there would be twenty-seven. If each of these twenty-seven passed this along to three of their friends, there would be eighty-one. If these eighty-one women talked to three others, in just four steps—there would be 243. In nineteen steps—three to the nineteenth power—this idea would reach over a billion women (1,162, 261,467).

This is how geometrical progression works, this is how a virus spreads and becomes an epidemic, this how consciousness-raising groups became the women's movement. This is how a call to gather the women could bring women together in villages, cities, in corporations, in governing bodies everywhere, to do whatever it is that is their assignment from Mother.

Full equality between men and women is one of the most important prerequisites for peace. Without full equality, there is injustice and the promotion of harmful attitudes in boys and men which is carried from the schoolyard and family to the workplace, to political life, and ultimately to international relations. Without full equality, qualities associated with women are suppressed in boys and men.

I believe that a 5th United Nations–sponsored international women's conference would be a giant step toward a tipping point—in terms of numbers attending, connections made, and a ripple effect. Only a U.N.–sponsored conference would allow women to attend who otherwise would not be able to get visas and support from their countries. Women would share information about what has worked, they would find role models, mentors, and allies. They would learn from each other and connect with kindred spirits.

The principle that women's rights are human rights and the Twelve Critical Areas of Concern that constitute the Beijing Platform for Action were the accomplishments of the 4th U.N. International Conference on Women in Beijing in 1995. Half the NGOs that exist now were not in existence in 1995. Each of the Twelve Areas could be a conference within the conference, bringing together those who are working on finding solutions for similar needs.

For there to be a 5th women's conference, we need to create grassroots advocacy for this in the countries that belong to the United Nations and at the UN Commission on the Status of Women meetings. This is the reason for the website *www.5wwc.org* and big blue 5WWC buttons, for 5th Women's World Conference.

On November 21, 2007, the General Assembly of the United Nations passed a draft resolution on The Right to Food by a recorded vote of 176 in favor to 1 against (the United States) stating that it is intolerable that more than 6 million children still die every year from hunger-related illness before their 5th birthday, further noting that women and girls are disproportionately affected by hunger, food insecurity, and poverty, partly due to gender inequality and discrimination, and that in many countries, girls are twice as likely as boys to die from malnutrition and preventable childhood diseases—it is estimated that almost twice as many women as men suffer from malnutrition.

Women in Columbus, Ohio, started a grassroots organization that is Mother-inspired in 2006. Through word of mouth and their website *www.standingwomen.org,* they encourage women to stand together on Mother's Day and affirm:

> We are standing for the world's children and grandchildren and for the seven generations beyond them.
>
> We dream of a world where all of our children have safe drinking water, clean air to breathe, and enough food to eat.
>
> A world where they have access to a basic education to develop their minds and healthcare to nurture their growing bodies.

A world where they have a warm, safe, and loving place to call home.

A world where they don't live in fear of violence—in their home, in their neighborhood, in their school, or in their world.

This is the world of which we dream.

This is the cause for which we stand.

This is Mother's agenda.

Jean Shinoda Bolen, M.D.
January 2008

INTRODUCTION

"Gather the women" is a message to her daughters from Mother Earth, Mother Goddess, Mother archetype. The words evoke an intuitive recognition, a wisdom whose time has come. It is a call from the Sacred Feminine to bring the feminine principle into consciousness. It is time to "gather the women"—for only when women are *strong together* can women be fiercely protective of what we love. Only then, will children be safe and peace a real possibility.

When I first heard the phrase "gather the women," the words struck a deep emotional chord. I am one for whom the message was meant, as you may be also. It is an urgent message from Mother to Her daughters—that will not be heard by women who are allied with patriarchy, whose identities and value come through their relationships with men and male institutions. Women who respond have a sense of sisterhood with other women and react with maternal concern to pain and suffering, especially toward those who are vulnerable and powerless.

There is a collective bigger story about being a woman in the twenty-first century, when the fate of the Earth and all life upon it is at risk. Coinciding with this, here we are, women who have been the beneficiaries of education, resources, reproductive choice, travel opportunities, the Internet, and a longer life expectancy than women have ever had in human history.

Twice before, American women have changed their world and been an influence on the world through collective action. The first was called "the women's suffragette movement." Political equality—the right to vote was the goal. The second has been simply called "the women's movement." Social, personal, and economic equality were the issues and goals.

I believe a third movement is stirring below the surface of collective consciousness and is gathering momentum.

This third time may become "the women's peace movement." The goal is to stop violence by involving women in prevention of violence, resolution of conflicts, and restoration of peace. Domestic violence, school ground violence, street violence, terrorism, and wars have the same origins in the need to dominate and to be predator instead of prey. Until women collectively become involved in creating a culture of peace to stop violence begetting violence in the human family, women and children will continue to be the primary casualties.

When people find themselves at a crossroads or in a crisis, to move forward toward health, reconciliation, and life, the challenge is to let go of an outmoded attitude, idea, or perception. Individually or collectively, a shift has to take place, a tipping point is reached, and then the phenomenon of "There is nothing so powerful as an idea whose time has come" kicks in.

Gather the Women

In January 2003, I was the recipient of the Woman of Vision and Action Award, gave a talk at a Friday evening banquet, and stayed overnight. At breakfast the next morning, I heard of "Gather the Women" for the first time. This was an Internet project inviting women to create gatherings on or about March 8, 2003: International Women's Day.

I felt the power of the words "gather the women" as soon as I heard them.

On checking my calendar, I saw that I would be in Ireland at a Jungian conference on that day. At the conference, I spoke about the Gather the Women project and asked if anyone were inspired by the idea to organize something. The evocative power of the three words had an effect. There were volunteers, and a ritual was planned for early in the morning of March 8. All those interested were invited to assemble outside the hotel, which was on the edge of Galway Bay.

A storm arose during the night that had not abated by morning: the wind was fierce, the waves of the bay now had whitecaps, and there was a driving rain. I wondered if anyone would go out in this weather, but having instigated the ritual and wanting to support the women who had planned it, at the appointed time I ventured out—my head down into the wind and rain, toward the huge rock that emerged like a round breast close to the shoreline, around which we were to meet. Others were hunkered down when I got there, and more would emerge in ones and twos from the hotel, until there were perhaps as many as twenty-five of us—four or five of whom were men—gathered around the rock. A song was begun, but as sound was taken up by the wind, only fragments were heard. We each had found a stone on the rocky shoreline that we placed on the huge mother stone with our intentions or prayers, none of which could be heard by anyone else.

This experience supported my intuition that the words *Gather the women* would have an evocative power on others as well. That men also came was an indication that there are men who recognize the need for women to take a lead and will be there to support what we do, even when storms arise.

When I returned home, I learned that our huddle of wind-blown and drenched people who celebrated International Women's Day in Ireland were one of 405 gatherings in twenty-three countries and thirty-eight states who registered their event.

If you heed the message *gather the women*, the first step might be a discussion with friends, or an invitation to them to form a circle with a spiritual center, or you may have an idea for the next International Women's Day. Energy of women together is generated by a mix of love, outrage, ideas, comments, infectious laughter, and a desire to make a difference.

Grassroots

Each gathering was organized at the grassroots level. Grassroots is a descriptive adjective that refers to something that grows from the bottom up through the concerned efforts of ordinary people. Something neither ordered nor organized from above, and as humble, unremarkable, and unnoticed as very small emerging clumps or individual blades of grass might be in an immense field. Women with concern for humanity and the environment carry within them seeds of compassionate activism. The response to this simple invitation to do something expressive was a small indication that women all over the world share concerns and are connected.

A wider perspective came from being at the United Nations when the Commission on the Status of Women meets. These annual meetings, held the first two weeks in March, bring together an international community of women in nongovernmental organizations (NGOs) that focus on women. A series of events that began when I wrote a book called *The Millionth Circle* led to me being at the United Nations with an organization of the same name. I had my consciousness raised by the readily available information, which is virtually ignored by American mainstream media. Reality on the ground comes through hearing about the scope of preventable suffering that affects women and children, coupled with indifference and exploitation by those in power. I was inspired by women who are on the front lines and in back offices, making a difference.

And, I learned of the UN resolutions and agreements that are already in place, which if taken seriously by the governments that signed them provide concrete steps toward ending violence and achieving peace. In the United States, legislation has been proposed for a Department of Peace, the adoption of which would be a significant step toward this same goal.

Millionth Circle

The seed idea for this organization was *The Millionth Circle: How to Change Ourselves and the World—An Essential Guide to Women's Circles.* "The millionth circle" is a metaphor for the circle that, added to the rest, brings about a critical mass that ushers in a new era. The idea of such an organization was germinated in Geneva by Elly Pradervand and Peggy Sebera. I first heard about this from Peggy when she returned from Geneva and called to ask if the name "millionth circle" could be used and if I might come to a meeting in Northern California to explore the idea of forming an organization. Twenty women who had worked with circles and felt an affinity to the idea came to discuss the possibility, and after a second meeting at Mother Tree retreat center, Millionth Circle—the organization—formed in 2001. Through a series of gatherings held the following year in New York State, Glastonbury and London, England, in Wales, and Findhorn and Iona, Scotland, links were made with women in those areas and others who came from Africa, South America, Europe, and India.

In the process of forming any new activist organization, there is a need to clarify the intentions and to model the principles. A small working circle crafted an intention statement out of the discussion and sent it out via e-mail for comment. The result was a shared vision, which included involvement with the United Nations:

> Circles encourage connection and cooperation among their members and inspire compassionate solutions to individual, community and world problems. We believe that circles support each member to find her or his own voice and to live more courageously. We intend to seed and nurture circles, wherever possible, in order to cultivate equality, sustainable livelihoods, preservation of the Earth and peace for all. We intend to bring the circle process into United Nations accredited nongovernmental organizations and the 5th UN World Conference on Women, and to connect circles so they may know themselves as a part of a larger movement to shift consciousness in the world. (From *www.millionthcircle.org*)

Every circle that considers themselves part of the millionth circle vision is linked through their intentions. PeaceXPeace (see Chapter Resources) links women's circles in the United States directly with circles in other countries, focusing especially where war has taken a toll. In Europe, *The Millionth Circle* inspired the Women's World Summit Foundation to form Circles of Compassion for men and women, and November 2 was designated as "The World Day of Circles of Compassion" as part of the millionth circle vision. There are probably thousands of organizations and millions of people who are unknowingly affiliated in their hearts. There are also countless men who feel that men cannot bring peace to the world and hope that women will somehow get the message and do so.

Save the World

Women are experienced in looking after the children and fragile elders, cleaning up, setting the house in order, being frugal with resources, putting food on the table, maintaining peace in the family, and staying on good terms with the neighbors. These are the same tasks that need doing in the community, and on a national and planetary scale. The world needs what women can do. The world needs "mother" to set things right in our unbalanced world.

Any woman can be an influence where she is. If you are in a circle that supports what you are doing, all the better. When women gather together, sharing stories is what we naturally do. This is how we learn and find encouragement, allies, and ideas. The largest gatherings—with the greatest influence on the women who take part and on the world—are the women's world conferences under the auspices of the United Nations.

There was a widely held and mistaken assumption that there would be a fifth international women's conference ten years after the last conference in Beijing, which would be 2005.

As of 2005, there was nothing even on the drawing board. A world conference of women by the end of the first decade of the twenty-first century can happen, but only if women in sufficient numbers make their wishes known to their elected decision makers and representatives to the United Nations. With the communication power of the Internet, this would be the most influential gathering of women ever held.

To the Reader

The urgent message from Mother is a call that can be heard and heeded anywhere on the planet. Wherever women with a sense of sisterhood and maternal concern gather, the message will be received. My words were written for you, if you need the words to go with feelings you already have, if you need support to believe that you can do something, or if you need inspiration or a strong nudge to act on the "assignment" that you know is yours to do.

The first three chapters are intended to bring you up to speed. The fourth chapter describes qualities that women have as a gender that would best serve humanity, *now*. The last three chapters will inform and inspire you and tell you how women individually and together are the antidote to the state of mind that has put us all and Mother Earth in harm's way.

Disclaimers may be helpful at the beginning. This is not about replacing patriarchy with matriarchy. I do know that women can be as power oriented and as unempathic as I am saying men as a gender learn to be, and that there are lots of men who are nurturing and empathic, as women are supposed to be. I am well aware of complexities and individual differences, which *Goddesses in Everywoman* and *Gods in Everyman* specifically addressed as archetypally based. Having said this, the position I take here is that women as a gender—as a whole, not every woman, but women generally—have a wisdom that is needed. It is time to gather the women and save the world.

1

MOTHER'S DAY

The original Mother's Day Proclamation, written by Julia Ward Howe in 1870, was not a commercial idea created to sell cards, flowers, or candy. It was a proposal to bring women of all nationalities together to bring peace to humanity. Howe had seen the horrors, devastation, and the aftermath of the American Civil War and saw war rise again, this time in Europe with the Franco-Prussian War.

This first Mother's Day Proclamation was a call to gather the women. It was directed to women to add their voice to "the voice of a devastated Earth" and called for women to take counsel with each other to find the means to bring peace to the world. The sentiments in the proclamation express what women the world over have felt since wars began. Now, at the beginning of the twenty-first century, it may be truly possible to bring this intention to fruition. Since the second half of the twentieth century, there has been a significant shift in the status and influence of women in the world, as well as an urgent necessity to find a means to end the threat of war, with nuclear weapons poised for use. Matthew Arnold predicted in the nineteenth century, "If ever there comes a time when the women of the world come together purely and simply for the benefit of [hu]mankind, it will be a force such as the world has never seen." Empowered maternal concern is an untapped feminine force that the world needs to balance and transform aggression.

The groundwork for women coming together to be such a force was done by the women's movement: women in the 1960s and '70s who opened doors that the Baby Boomer generation came through in great numbers. In a matter of decades, women had opportunities and positions in the world that women had never had before. The second element that would make this possible is the communication technology that developed during these same years, so that information and images are now sent almost instantly all over the world. Women can meet, discuss ideas, and make plans through e-mails, arrange for translations, have conference calls, and forward news to all their friends with a key stroke. The third element is the emergence into consciousness collectively that it is up to women to change the world.

The original Mother's Day Proclamation was an expression of the concern that women can have for each other's children, the importance of expressing grief and sorrow, and then getting on with finding ways to bring about peace.

> Arise, then, women of this day! Arise all women who have hearts, whether our baptism be that of water or of tears!
>
> Say firmly: "We will not have great questions decided by irrelevant agencies. Our husbands shall not come to us, reeking with carnage, for caresses and applause. Our sons shall not be taken from us to unlearn all that we have been able to teach them of charity, mercy and patience. We women of one country will be too tender of those of another country to allow our sons to be trained to injure theirs."
>
> From the bosom of the devastated earth a voice goes up with our own. It says "Disarm, Disarm! The sword of murder is not the balance of justice." Blood does not

wipe our dishonor nor violence indicate possession. As men have often forsaken the plow and the anvil at the summons of war, let women now leave all that may be left of home for a great and earnest day of counsel. Let them meet first, as women, to bewail and commemorate the dead. Let them then solemnly take counsel with each other as to the means whereby the great human family can live in peace, each bearing after their own time the sacred impress, not of Caesar, but of God.

In the name of womanhood and of humanity, I earnestly ask that a general congress of women without limit of nationality may be appointed and held at some place deemed most convenient and at the earliest period consistent with its objects, to promote the alliance of the different nationalities, the amicable settlement of international questions, the great and general interests of peace.

Julia Ward Howe, Boston, 1870

One hundred and thirty-five years later, on December 26, 2004, Amalia Avila González, mother of Marine Lance Cpl. Victor González, flew more than nineteen hours from San Francisco to Amman, Jordan. Victor González, nineteen, was killed in combat in Iraq, barely a month after he'd arrived. During eight days in Jordan, Amalia Avila González met Iraqi refugees, including mothers like her who had lost a son or a relative in the war. The delegates from Global Exchange and Code Pink, the two groups that organized the trip, traveled with translators, but González said she understood what they felt because of their common bond as mothers: "They cried."

Motherhood, Mother Archetype

The Global Peace Initiative of Women Religious and Spiritual Leaders in Geneva, which I attended in 2002, was an historic first meeting of several hundred delegates. This was an unprecedented international meeting at the beginning of the twenty-first century, sponsored by the United Nations, that recognized the untapped potential of women spiritual and religious leaders as a necessary force for peace. At this conference, the Gandhi-King Peace Award (previously given to Kofi Annan, Nelson Mandela, and Jane Goodall) was given to Amma, who is best known in the West as the hugging guru. In her acceptance speech, this spiritual leader from India said, "With the power of motherhood within her, a woman can influence the entire world. The love of awakened motherhood is a love and compassion felt not only toward one's own children, but toward all people, animals and plants, rocks and rivers—a love extended to all beings."

Amma's definition of motherhood was archetypal and eloquent: "It is not restricted to women who have given birth; it is a principle inherent in both women and men. It is an attitude of the mind. It is love—and that love is the very breath of life."

Mother archetype, maternal concern, and Amma's description of motherhood are interchangeable. Until maternal concern has a strong voice—that is heeded—on matters of peace and security, the agenda for the world will not change: it is about control and acquisition of power, which are the basic patriarchal goals. The specific items on the agenda change, but the motivation remains. Power-oriented leaders determine what matters; men follow; women obey the men and tend to household and children. Patriarchy considers this the natural order and war an effective or necessary means to gain control.

Different Perspectives on War: Gender Differences

Six months after the Women's Global Peace Initiative in Geneva, the president of the United States decided that the danger that Saddam Hussein posed was sufficient to necessitate invasion of Iraq. When the invasion began, there were journalists embedded with the military and television crews on the ground. There were maps with arrows marking the unimpeded progress of the invasion, which had been code-named "Operation Shock and Awe." As a generality (by this I mean that what I am saying applies to most men and most women, but definitely not to all), there was a decided gender gap in response to the invasion, even among women who believed that it was necessary.

I think it would be fair to say that men were impressed, interested in seeing and hearing about the equipment and the strategy. In bars, large screen televisions were turned on as they normally are to football. The experience was, in fact, very much analogous to watching a sport. The arrows marking troop movements were like those that are used to demonstrate successful plays: who carried the ball, who ran interference, how many yards were gained. That our team is bigger and stronger and has a decided advantage is all the more reason to cheer, as our team moves ahead and scores. Only war is not a game, even when it is on screen.

Most women were also following what was happening on TV during the first days of the invasion, more with concern than admiration. For mothers, an 18-to-24-year-old son or daughter is not much more than a kid. It was easy to imagine one's own in harm's way. It was also easy to think that innocent people were going to be hurt. When the nighttime sky was illuminated by bomb blasts, it crossed our minds how hard it must be to live there and how terrified the children would be.

The weekend of the invasion coincided with a Millionth Circle gathering in the Bay Area. I had several friends staying with me who had come for the meeting. We watched television together and were appalled that this was happening. The only lightness came from appreciative comments we made about David Bloom, our favorite embedded journalist. In just a matter of weeks, I learned that he had died. Within the year, what we older mothers dreaded came to pass: every day there were photographs of young people killed in Iraq with their names, rank, age, and home towns. Unmentioned were the six or ten others wounded, many horribly so, for every soldier who was killed, and the silent damage that will surface as traumatic stress disorders when the troops come home. Unnewsworthy were the numbers of casualties in the civilian population.

There are gender differences. Psychologist Simon Baron-Cohen says that the essential difference is that women are natural empathizers, while men are better at systemizing. Most women who were tested agreed with statements such as, "I get upset if I see people suffering on news programs," or "It upsets me to see an animal in pain," or "Friends usually talk to me about their problems," or "I can usually appreciate the other person's point of view, even if I don't agree with it." Men who are tested usually do not agree that this is so for them.

When the agenda for the world is determined by men, it means that decisions and actions that affect the planet, its people, and all life upon the Earth are made by the gender that most likely does not know or care about what others are feeling, experiencing, or suffering. Until women are really involved in what goes on in the world, essential information and crucial concerns are not brought to the table.

What if it were up to mothers to make the decision to go to war? This was so for the Iroquois Confederacy, the people who are also called the Seneca Nations and who still maintain their sovereignty in the northeast United States. The elected Council of Clan Mothers were grandmothers, women whose own children were grown and who were beyond their childbearing years. They determined the priorities for the confederacy, including whether to go to war. If war was decided, the conduct of the war, including electing the war chief, would then go to the Men's Council, whose members had been nominated by the Council of Clan Mothers. Deliberations were not made in haste. The experience of the past seven generations and the effect upon seven generations to come is taken into consideration. A wise and sensible consideration, because war and its aftermath invite retaliation, retribution, and revenge for the past and may involve generations to come.

Maternal Concerns, Women's Rights, First Wave Feminism

Women want a world that is safe for children, one in which they do not live in fear themselves. It will never happen unless women as a gender become actively involved and full partners in determining the fate of the Earth and life upon it. Toward this end, every effort to empower and educate women counts, as well as every neighborhood and school made safer. For peace to become a reality, women have to gather together, learn from each other, and then work with men toward ending violence as a means of winning arguments or gaining power—in households or in the world. In recent years, American women have specifically mobilized maternal protective instincts and sister-bonds effectively. It has resulted in MADD—Mother's Against Drunk Driving, which has affected laws, sentencing, and created the designated driver. The Million Mom March, a demonstration for gun control, was started by Donna Dees-Thomases after a gunman randomly shot a group of school children. It was an appeal to gather outside the White House on Mother's Day 2000 to demand the passage of gun control legislation. Seven hundred and fifty thousand demonstrators showed up, while simultaneously sixty marches took place across the country. Protecting innocents, *enough is enough!* outrage, and indifference from the powers that be are making activists out of mothers.

To gain a voice and have an influence in the world, women have had to first stand together to overcome ridicule and disregard. Individually and together, women have had to face threats of violence against them and been willing to be arrested in order to gain the right to vote (suffrage). It took seventy years of political effort for women to vote in the United States, achieved through a constitutional amendment in 1920. In Britain and Ireland, women over thirty gained the vote in 1918, by an act of Parliament. "Suffrag*ettes*," now a respectable word, was initially derisive—used to minimize women Suffragists. Their efforts were denounced from pulpits as being against God's will. When they marched in the streets, they were spit upon, laughed at, and some were arrested. Of those who were jailed, numbers of them were beaten. It is easy to forget that rights women take for granted now are historically very recent and were gained for us by women who were strong and courageous together. The right to own property, the right to keep money earned, the right to marry without a father's or a father surrogate's permission, the right to be educated, and the repeal of laws such as one that gave a husband the right to discipline his wife with a stick, as long as it was no thicker than his thumb, all occurred in the context of women seeking the right to vote. This was the first wave of feminism.

Second Wave: The Women's Movement

The second wave was the women's movement. It brought about social, economic, personal, and political changes, and defined new rights. It had its beginnings in the mid-1960s with Betty Friedan's *The Feminine Mystique* and President John F. Kennedy's 1963 *Report on the Status of Women,* which documented women's economic inequalities.

The women's movement began in the minds of women who began talking together about their own lives and examining the premise that they were inferior to men and the laws and common practices that supported this. Consciousness-raising groups arose spontaneously whenever one or more women decided to call friends together. Ideas are infectious, and ideas of inequality and oppression became understood as patriarchy and spread rapidly through the collective consciousness of women. Each consciousness-raising group generated energy, and both contributed to and drew from the women's movement.

In these circles, women shared personal stories, saw common themes, and became aware of sexism. With the support of each other, individual women challenged stereotypes, defined themselves, spoke truth to power, and strove for egalitarian personal relationships with men. They raised each others' awareness of what needed to be changed in society and in personal situations. The ringing theme in the '70s, the decade of the women's movement, was "the personal is political." Women had found out that their personal lives and politics—power inequality—in the economic, social, and political spheres were related. Relationships, stereotypes, and laws changed as a result, and these changes rippled out and were an influence in the world.

Third Wave: The Women's Peace Movement

I believe that the third wave of feminism is taking shape, much as waves themselves form in the ocean. They arise from deep below the surface, away and out of sight just as thoughts, intuitions, and feelings arise in the psyches of individual women and gain momentum as they spread to others. New ideas become a movement when the force and energy behind them overcomes resistance to change. I believe the third wave of feminism will be a women's peace movement that is growing out of the recognition that *only* when women and children are safe from violence, deprivation, and abuse will the cycle of violence begetting violence, which underlies terrorism and wars, end. Compassion, spirituality, the desire and necessity for peace, and maternal concern, combined with feminism is the force that can save the world.

The first Women's International League for Peace and Freedom conference, held in 1915 in the Hague, The Netherlands, was the equivalent of the first Women's Rights Conference held in 1848 at Seneca Falls, New York, which began the suffragette movement in the United States (which took until the next century to achieve). In 1915, during World War I, 1,300 women from countries at war against each other and from neutral countries attended. Their vision was similar to that expressed in the original Mother's Day Proclamation. Their proposals for a lasting peace are still relevant, as is the active organization that grew out of this conference.

The women's movement raised consciousness about patriarchy and the use and abuse of power that can be applied to understanding the causes and effects of war. The psychology of unequal relationships, where one person has power over the other and can harass, humiliate, rape, control, or intimidate the other, often can be applied to conflicts between nations. War is on a large scale, similar to domestic violence for children: it is serial and chronic traumatic stress. Just as the second wave of feminism grew out of the first, a third women's peace movement can grow out of the women's movement.

Traumatized Children and Oppressors

The concerns of mothers—to make the world safe for children—would move the world toward peace and sustainability for everyone. Most oppressors who seek to intimidate or exercise control over others felt humiliated and were often traumatized as children or adults by people who had power over them.

Anxiety begins in the womb of the pregnant mother who is terrified by the violence around her or fearful for herself and her unborn child. Cortisol, the stress hormone, which goes up in traumatized pregnant women, crosses into the placenta and affects the brain of the fetus. These mothers give birth to infants that are often premature and small, who grow into children with a predisposition toward poor impulse control, inattention, and learning and behavior problems. These propensities would be made worse by witnessing violence and being a target of anger and abuse themselves. Violence does beget violence.

Older children dominate younger ones, boys abuse girls, a dominator pattern results. Basic trust develops in children who are nourished and nurtured and have mothers who can respond to their distress and needs. In contrast, children who live in war zones do not feel safe, are startled by loud noises, by gunfire or explosives, or angry or terrified voices. A bad neighborhood with drive-by shootings, or households in which domestic violence erupts and women and children are hurt, are war zones for those who live there. In all such situations, children's needs are ignored; they are in harm's way, and feel abandoned by adults who go away for any reason. Without an adult or a society to protect them, children are vulnerable to whatever bad happens. Boys wait their turn to be men with the upper hand; girls become acculturated to becoming powerless women.

Amnesty International's Stop Violence Against Women: It's in Our Hands calls the statistics on violence against women a human rights catastrophe: at least one out of every three women worldwide has been beaten, coerced into sex, or otherwise abused in her lifetime. Abuse of pregnant women by their male partners is not uncommon. Usually the abuser is a member of her own family or someone known to her. A common trigger for violence is refusing sex.

Domestic violence is the major cause of death and disability for women aged sixteen to forty-four and accounts for more death and ill health than cancer or traffic accidents. In the United States, women accounted for 85 percent of the victims of domestic violence. Up to 70 percent of female murder victims are killed by their male partners. Besides domestic violence, Amnesty International describes violence toward women in the community and by the state, which includes acts committed or condoned by police, prison guards, border guards, and so on, as well as rape by armed forces during conflict, and against refugee women or women held in custody.

Disempowered and fearful mothers cannot protect their children no matter how much they may love them. To an infant or young child, mother is all powerful. She is the source of food and comfort, of approval or punishment. Adults are giants compared to children's physically small selves. Then, if a mother (who herself could be a child bride in many traditional cultures) cannot protect them or provide for them, children feel deeply betrayed, not just by the mother, but by the world. Powerless mothers instill mistrust and devaluation of women in boys and girls.

Safe and Empowered Girls and Women

When girls are educated, literate, and knowledgeable about nutrition and spacing their children and have positive role models, they marry later and have fewer children, and those children are born healthier. As a result of individual women making choices in their best interest and in the interest of healthy, wanted children, the planetary problem of overpopulation also is eased.

Ingrained in patriarchy is that women belong to men, and that male potency is reflected in the number of children they father. The more patriarchal the family, the religion, and culture, the younger when married, less educated, and less independent the women. Women's sexuality and childbearing is then in the service of men. *Roe v. Wade* gave every woman, in principle, the right to determine whether she will bear children or not. This right has been undermined and efforts to reverse this law continue. When this choice is up to her, it fundamentally undermines the patriarchal principle that men, individually or as religion or government, have the right to control women's bodies. Without access to birth control and reproductive choice, traumatized women who cannot refuse sex are also forced into bearing children, who will also suffer the consequences.

There is a blueprint to begin construction of a world in which women would be safe from violence, exploitation, and discrimination, could look after the well-being of their children, and have a voice in all areas including the environment. It is spelled out in the *Beijing Declaration and Platform for Action,* which was adopted by the United Nations Fourth International Conference on Women in Beijing in 1995. It names twelve critical areas of concern, with specific steps toward remedying them. Signing a document such as this was a major symbolic step that required overcoming a lot of resistance.

Actualization is what really counts, however, and the truth is—in this patriarchal world—concern for women and girls does not rank very high in importance.

This lack of concern is horrifyingly reflected in the trafficking of women and children, mostly girls, which is a huge international business. Women in Third World countries are lured by promises of marriage or work, are raped and beaten until they are docile and cooperative, are transported from country to country, to be used and abused sexually. Little girls are in this same pipeline, bought or kidnapped to provide for the sexual appetites of men who pay well for pubescent girls or even younger ones. Fetching less money but still a profitable commodity are women and children who are sold into domestic or factory slavery. Very little if anything has been done about this trade, even though reports to the United Nations estimate that this involves more than one million women and girls every year. Only after a missing twelve-year-old Swedish boy was reportedly seen at a clinic after the 2004 tsunami did Americans hear about the trafficking in children or learn that children who had been separated from parents or were now orphans could be prey for profit.

For women, peace is not just the absence of war, but safety and security for their children and grandchildren and freedom from terrorists of all kinds, including those who represent their own government or commit domestic violence upon their families. A mother with global consciousness knows that it is not only her children and grandchildren or the children in her community, or even in her country, but everybody's children, everywhere, who are deserving of a good and safe life.

How different reality is! The United Nations Children's Fund in its 2005 Report "Children Under Threat," says that more than half of the world's children of more than one billion suffer extreme deprivation because of war, poverty, and HIV/AIDS. The world is small. A deprived and abused child soon becomes an adult, and, as an enraged adult with the power to harm others, may do just that.

Mother Power of Women Together

The dormant power of women together is the untapped resource needed by humanity and by the planet. Only when mothers are strong in spirit, mind, and body will it become possible for children to be wanted, nourished, and secure. It would then be possible within several generations to bring about an evolutionary change in relationships between men and women for the benefit of everyone.

To exercise rights or claim rights not given to us in order to look after our home planet, the human family, and those we share the Earth with *is* women's work, best done when it is done together.

2

MOTHER EARTH/ MOTHER GODDESS

In classical Greek mythology, Gaia was the Earth Goddess. She was the first to take form out of chaos. She then gave birth to the sky, mountains, rivers, oceans, and all living things on the planet. The words *matter, matrix,* and *material* come from *mater*—Latin for "mother." To imagine divinity as female was a natural consequence of observation. Anyone could see that new life emerges out of female bodies and that all living newborns survive only if they are nourished and protected by their mothers. We refer to Mother Earth and Mother Nature easily and naturally. It was once just as easy and natural for people to worship the Mother Goddess, the Divine Feminine, and see women in her likeness.

Ancient people who worshiped the Goddess lived close to nature. I imagine that they felt awe and wonder that a baby could grow inside a woman and be born. I think that births are awesome, scary, and wonderful experiences. I loved delivering babies, in medical school under supervision, and then in internship when I took extra rotations on the obstetrical service. Birth is earthy, messy, and miraculous, which I think would likely be the reaction of people who retain the ability to feel awe and wonder. But when men looked upon women as inferiors, then anything women did that men could not, was defined as being like an animal and not like a goddess. The Sacred Feminine was no longer acknowledged.

Waves of invading warrior tribes with their sky gods conquered the people of the Goddess in Old Europe by overpowering them. As the invaders and indigenous people intermingled, goddesses were incorporated into the patriarchal religion as consorts and daughters. The *Iliad, Odyssey,* and poems attributed to Homer tell us about the attributes and myths of the Greek divinities. Hesiod's *Theogony* tells us their cosmology. Western civilization, beginning with the ancient Greeks, is patriarchal, based on male power. Yet, even in their cosmology, before Zeus with his thunderbolts ruled from Mount Olympus, there was Gaia. Goddess preceded god.

When God Was a Woman

It wasn't until the twentieth century that archeology became a profession, and not until the second half of the twentieth century that archeologists could get past the androcentric assumption that divinity was always male to understand the meaning of the artifacts and abundant female statuary they were finding. Archeologist Marija Gimbutas cataloged and analyzed hundreds of archeological sites, which she designated as Old Europe, which flourished and developed between 7000 and 3500 BCE. These were not primitive people or barbarians. They lived in locations chosen for their beautiful setting, good water and soil, pasture and farmland. They cultivated wheat, barley, peas, and other legumes, and bred all the domesticated animals still found in the Balkans today, except for the horse. There was stability, development of crafts, pottery, copper metallurgy, creation of jewelry and statuary; there was trade and communication, sailing, a rudimentary script, and evidence that these were people who loved and valued art. Remarkably absent were fortifications or thrusting weapons. Archeologist James Mellart excavated thriving communities in southern Turkey, Catal Hüyük and Hacilar, dated between 7000 and 5000 BCE and also found no evidence of warfare or central authority.

From archeological evidence, the Neolithic farming people of Old Europe and Turkey were goddess worshippers who had maintained stable, peaceful communities that lasted for 2,000 years in Turkey and 3,500 years in Old Europe. In *The Chalice and the Blade,* Riane Eisler synthesizes available evidence into an interpretation of history that takes into account the existence of a harmonious and peaceful age that preceded patriarchy. She notes that nineteenth- and twentieth-century scholars, on realizing that people in Neolithic times worshiped the Goddess, jumped to an erroneous conclusion that if society wasn't patriarchal, it must be matriarchal—meaning that if men did not dominate women, women must have dominated men; or that if women didn't dominate, then male dominance must have always been the norm. Eisler speculates from the evidence that this was not an either/or situation. Instead, neither gender was subordinate to the other. Eisler proposes that the power symbolized by the Chalice was *actualization power,* as distinguished from the *dominator power* of the Blade. She documents the triumph of patriarchy as a triumph of a power-oriented, death-focused religion that divinely sanctioned killing in war.

The premise that before there was God, there was Goddess is supported by archeology and by mythology. When nomadic warrior people from the North, beyond the steppes, swept down into goddess territory with their sky gods and weapons, they subdued the unarmed people and unfortified communities. Gimbutas, in *Goddesses and Gods of Old Europe*, called these warrior people "Kurgans," with "Indo-Europeans" the alternative name, and noted that there were successive waves of invasions into the Greek peninsula, and that over time the nomads became settlers, mating with the indigenous Goddess-worshiping people, incorporating the Goddess into their beliefs, to eventually become the ancient Greeks.

What Happened to the Goddess?

Leonard Shlain in *The Alphabet Versus the Goddess* proposes that literacy brought an end to the Goddess, caused the decline of women's social and political status, and ushered in patriarchy and misogyny by reinforcing left-brain dominance, with linear thinking now valued over feelings and intuition, word valued over image, and hierarchy the natural order.

Writing *Goddesses in Everywoman,* which was published more than twenty years ago, provided me with the impetus to read about what was known about the history and archeology of the Goddess, though I was writing about goddess archetypes and drawing most upon mythology and psychology. Through archeology, I could appreciate that these myths and archetypal patterns evolved within patriarchy.

Shlain provides an explanation that I think further explains how patriarchy came to dominate. I don't think an either/or choice needs to be made to explain what happened to the Goddess and to women under patriarchy. With the alphabet, patriarchy, which had previously been based upon male physical strength, weapons, and military strategy, gained authority based upon written words. Words now supported and justified holding power and religious authority.

When Shlain pointed out that there is no word for "goddess" in Hebrew, I had an *aha!* insight. In Sunday school and in college, when I read about the sin of worshiping "false gods," I had always thought they were males, never goddesses. Nor did it occur to me that the statues or graven images that were abominations in God's sight might have been of a young maiden, a maternal or sensual figure, or old woman, each a representation of the one Goddess in her three forms. It was an easy mental jump from this to the realization that Canaan, the promised land of milk and honey, was an agricultural, peaceful land of abundance, with cities and vineyards and art, whose major divinity was the Goddess. It was a war against people who had done nothing hostile, a war sanctified by the God that the three major monotheistic religions worship. I sought passages in the Old Testament to read God's instructions to his invaders. In Deuteronomy 7, the Israelites were directed to defeat and utterly destroy the people, to show no mercy, to not intermarry or enter any agreements. They were to destroy their altars, cut down their sacred groves, and burn their graven images.

Shlain casts new light upon the Ten Commandments, especially the first two. The First Commandment is, "I am the Lord thy God. Thou shalt have no other gods before me." In the context of ancient times, this was a God unlike all deities before him. He had no need of a wife or consort, there was no sacred marriage. God created through his word. The Second Commandment, which almost no one remembers, is, "Thou shalt not make unto thee any graven images, or any likeness of any thing that is in heaven above, or that is in the earth beneath, or that is in the water under the earth" (Exodus 20:4). Shlain points out that this proscription against making images is repeated throughout the Torah (the first five books of the Old Testament). There are to be no illustrations, no colorful drawings, no painting or statues, no representational art.

The Taliban, a militant, fundamentalist, Islamic sect, demolished two colossal statues of Buddha in 2001 that had been carved a thousand years before into the sandstone cliffs of Bamiyan in Afghanistan. The Taliban claimed that the Buddhas violated the Islamic prohibition against sacred images and that they were false idols that must be destroyed. The Christians who knocked the noses and limbs off Greek statues and defaced Egyptian temples may have been similarly motivated. Protestants in their unadorned churches to this day often look unfavorably upon statues in Roman Catholic Churches, seeing them as idols.

The prohibition against representational art meant that people would be turned away from appreciating or—

contemplating beauty, much of which is in nature. Love and beauty, beauty in the eye of the beholder, eyes as the windows of the soul become lost connections. Shlain notes that the prohibition of art would support left-brain dominance. The need to prohibit beautiful images becomes unnecessary once people lose the ability to see or be touched by beauty, or be awed by nature, or to feel that any*thing* can be sacred. The photograph of Earth from outer space affects only those who still can be emotionally and spiritually moved by an image.

The Gaia Hypothesis

In the mid-1960s, James Lovelock, an atmospheric scientist, and Lynn Margulis, a microbiologist, formulated the Gaia Hypothesis, which proposes that the Earth is alive and that it functions much as our bodies do to maintain homeostasis. Our bodies are an environment in which the temperature and chemistry has to be kept within a very narrow range of fluctuation. To be alive and healthy requires very sensitive interaction among many systems—circulation, respiration, digestion, elimination, hormonal, and neurological systems, and so forth. Gaia—the planet Earth—maintains equally complex interactive systems involving the atmosphere, oceans, soil, and biosphere that also maintain an optimal physical and chemical environment for life. The Gaia hypothesis was startling because it proposed the idea that the Earth is a single, living entity; that she is alive.

We inhabitants of planet Earth are in a relationship to Gaia that I think of as like that of a fetus in the body of the mother. While in the uterus, we move in amniotic fluid, with all the nutrients we need to grow and thrive coming from our mother's body. In this womb space, we are provided for and protected, much as the atmosphere protects and provides for life on Earth.

Detractors of the Gaia Hypothesis noted that there is a very narrow band of life at the surface, and that below this the Earth is composed of inert material, mostly iron: how could the planet be alive when most of it is inert or dead? Proponents responded that the Earth is alive like a tree is alive.

I often walk in Muir Woods, under towering old-growth redwood trees, most hundreds of years old, some with life spans in the thousands. Among these giants, under their green branches, I am in a green cathedral. There is no question but that these trees are alive. And yet, like most of planet Earth, the wood in the trunk of the tree is not alive. There is only a rim of cells along the periphery of the trunk of a redwood tree that is living. Ninety-seven percent of it is not alive; the wood of the trunk and the bark of the tree are inert. Like the Earth, there is a thin layer of living organisms spread across the surface of the body of the tree. The bark, like the atmosphere, protects the living tissues of the tree and makes possible the exchange of carbon dioxide and oxygen. Trees in effect breathe in the carbon dioxide we exhale, and breathe out the oxygen we need to keep the atmosphere constant in composition; the rain forests are the lungs of the Earth.

Earth Seen from Outer Space: Portrait of Mother

In 1968, America successfully sent the first manned flights into space as part of the Apollo program. Inspired by President John F. Kennedy, the overall goal was to send men to the moon. Each launch furthered this goal, which was successfully achieved by the *Apollo 11* mission in 1969. The technology that led to this achievement was an expression of the masculine principle at its finest. It required objectivity, teamwork, leadership involving excellence of ideas and talent, and the ability to integrate many details and systems, in order to work out the problems.

To see the Earth from outer space was not an important goal of the Apollo program. Yet the photograph of Earth turned out to be far, far more significant than those of men in space suits on the moon. It initiated a quantum shift in human consciousness.

The photographs taken by the Apollo astronauts showed us an Earth that was beautiful. We gazed at an illuminated, mostly blue and white sphere surrounded by the black void of space. There were swirls of white clouds against the dark blue of oceans and the occasional brown of a continent.

This was an image that could touch the soul of anyone who was receptive to the beauty and meaning of what we were seeing for the first time. This was a photograph of our Mother the Earth, of Home. Not only could we see her beauty, but we could also realize her vulnerability. Seen against infinite space, how small and fragile Earth is!

The geometry of the photograph of Earth is of a circle within a square. This is the shape of a *mandala,* a Sanskrit word, the form for Tibetan sacred paintings, and according to C. G. Jung, a visual image and symbol for the Self, the archetype of meaning that we might call God, or Goddess. For many people, the image of Earth from outer space had the impact of an icon—an image imbued with sacred energy.

When poet and novelist Alice Walker gave the commencement address at her alma mater, Spelman College, in 1995, she described wondering what she had to give them, saying, "At last, it came to me. I can offer you the gift of my experiences." She spoke of her poems as the essence of what she learned, and offered them as medicine to use when needed:

> What can I give you to help you bless the day when you fully understand that the most basic fact that all patriarchal religions try to deny and to make people forget is that the Earth is our Mother and that She must be honored, in order for our days to be long on this planet? I give you: "We Have a Beautiful Mother."

> We have a beautiful
> mother
> Her hills
> are buffaloes
> Her buffaloes
> hills.

We have a beautiful
mother
Her oceans
are wombs
Her wombs
oceans.

We have a beautiful
mother
Her teeth
the white stones
at the edge
of the water
the summer
grasses
her plentiful
hair.

We have a beautiful
mother
Her green lap
immense
Her brown embrace
eternal
Her blue body
everything
we know.

Concern for Gaia and Maturity

The photograph of Mother Earth could only be taken by astronauts who were able to get far enough away to see our home planet from a distance. This is analogous to growing psychologically until we are mature enough to see our mothers as they are. Until we grow up, we have a self-centered relationship to our own mother. She is there to do for us, she is seen as it pleases us to see her, and not as separate from our needs and assumptions. When we finally are able to see our mother as a person and can love her as she is, we usually are mature enough to also realize that she may need us.

The photograph of Earth from outer space in 1968 recruited hearts and minds into a growing environmental movement. Greenpeace began campaigning against environmental degradation. Other new organizations were formed, such as the Green political parties that began in Europe. Old organizations such as the Sierra Club, founded in 1892 by John Muir, had a leap in membership and activism. The photograph of Mother Earth is without national boundaries. It is a statement that the people of the Earth share one environment, a rebuttal to the "us and them" mentality that thinks the United States can erect a wall around itself and not be affected by what happens elsewhere. The photograph of Gaia—Our Mother the Earth—and Rachel Carson's *Silent Spring*, published during this same decade, represent the two energies that mobilize environmentalists: love for the Earth and awareness that life on Earth is being destroyed.

In northern California, the logging of old growth redwood forests brought out activists who used words such as *Desecration!* and *Sacrilege!* These words convey outrage at the destruction of trees that are beautiful, ancient, and sacred. The corporate executives who gave the orders and the men whose cutting machines could bring down a tree in minutes see their value only in dollars. They do not feel that there is desecration in turning a tree that is more than a thousand years old into planks of lumber, nor take to heart a message from Greenpeace: "When the last tree is cut, the last river poisoned, and the last fish dead, we will discover that we can't eat money."

The Millennium Ecosystem Assessment (2005) conducted over four years by 1,300 experts from ninety-five countries, reported that humans had ruined approximately 60 percent of the Earth's ecological systems, and that this has resulted in a substantial and largely irreversible loss in the diversity of life on Earth. This study also said that reversing the degradation of ecosystems while continuing to use them to improve people's living standards was possible if radical changes were made in the way nature is treated at every level of decision-making, and if new ways of cooperation were developed by business, government, and civil society.

Love and Beauty

Seeing beauty, loving what is beautiful, and nurturing and sustaining it all go together. It is also the ability to sense or intuit potential beauty and, through love, encourage it into existence. Aphrodite was the Greek goddess of Love *and* Beauty, which acknowledges the archetypal connection of one with the other. When you fall in love with someone or something, you see the beauty that is there or potentially there. Maternal instinct is a falling-in-love experience. The young of all living creatures need mothers (and fathers with this archetype) to nurture and protect them. They are also universally cute: their eyes—windows of the soul—are large and soft, endearing to anyone with maternal instinct.

To respond to beauty and vulnerability is a human response that can be encouraged or suppressed. It is enhanced in women by hormones and role, and suppressed by ridicule in most male-dominated societies. Soft-heartedness is punished as unmanly, which means that human qualities such as empathy and compassion are socialized out of little boys, especially toward those who are not like themselves. Killing an animal is an initiation for many young males. On farms, a boy passes this test when he kills an animal raised for meat. In the woods, it is the excitement and accomplishment of bringing down game with a gun, and the praise of men, that overrides any misgivings that a boy might have.

Men who harvest endearing baby seals for their white pelts by bashing them cannot respond to their cuteness, or feel pity for their distressed and helpless mothers. Soldiers cannot let concern for collateral damage—the noncombatant casualties who are killed or maimed—get in the way of a military objective or an order. The importance of the profit margin—the bottom line—easily leads to downsizing if there is no concern for the impact of unemployment on people and families. The words—*harvest, collateral damage, downsizing*—are impersonal and mechanical. Words such as these are devoid of emotion, and help those responsible for decisions or actions that cause pain and suffering not to feel anything about what they do. When people or animals are only dispensable or profitable objects, they become just numbers. Language that uses accounting or mechanical terms when the subjects are living beings who will suffer creates emotional distance and numbs feelings of compassion. There is something very wrong with individuals who do not see beauty and can't mourn its destruction, or feel compassion and be affected by the suffering of others.

I believe that the socialization of very young boys, who are shamed for being soft-hearted, emotional, imaginative, or sensual, and are discouraged from interests and inclinations that men consider unmanly, contributes to one-sided development. The other contribution is fear and humiliation. Unprotected, neglected, and abused boys and young men grow up to emulate those who had power over them. Until mothers are able to protect both their boys and girls from abuse and bullying, boys cannot grow into whole people.

Women and Wholeness: A Gender Advantage

Wholeness is possible when human qualities, now usually designated as masculine or feminine, are seen as part of the spectrum for everyone. In the male world, intellectual development (rational thinking, data-based information, objectivity, the realm of the mind) is fostered and rewarded; emotional development (feeling, intuition, aesthetic appreciation, subjectivity, the realm of the heart), usually not. When both are important, both sides of the personality develop and both hemispheres of the brain are used. Since the women's movement, girls, especially, have had social support to develop both sides of themselves and have been freer to do so than boys. As a consequence, women as a gender are more likely to be whole people in this sense.

Universal education develops the rational side of the brain in both girls and boys, but social circumstances determine whether emotional intelligence or empathy will also be developed. Neurologically, "use it or lose it" occurs throughout life. The more connections there are between the two cerebral hemispheres, the greater the possibility of having thoughts and feelings, of having an experience that can be described objectively, and by words that are metaphoric and poetic. The more connections, the more myelinated fibers there are; the more fibers, the thicker the corpus callosum.

The corpus callosum is the structure that connects the two halves of our brain. In 1982, women were found to have a third more fibers in the corpus callosum than heterosexual men on postmortem research. The corpus callosum can now be seen in living subjects through MRI studies. In 1991, research was done at UCLA on 122 age-matched adults. Dramatic differences were found between men and women in the shape and width of regions of the corpus callosum. The size and width of the corpus callosum was significantly greater in women. Not surprising considering the multitasking women usually must do while paying attention to relationships, which requires considerable back-and-forth between the cerebral hemispheres.

The study also included twenty-four age-matched children (twelve paired girls and boys between the ages of two and sixteen). There were no significant corpus callosum differences between these boys and girls. I find myself speculating why this might be so. It could have to do with the egalitarian socialization and education in southern California. It could be that male and female brains start out more alike than different, and that differences arise when the emphasis is on one-sided development. At puberty when bodies are affected by hormones, it could be that brains might also change. Then again, puberty is when there is the most anxiety about being masculine enough or feminine enough, and conformity to stereotypes is the strongest.

Nature and nurture are involved in who we become, but living in the world that we do, by the time we are adults, there are gender differences: less so, the more egalitarian the culture; more so, the more authoritarian-patriarchal the culture. Countries (and families) in which the fundamentalists are in power oppress women and homosexual men. Where masculinity is defined as power and being in control, men fear feminine qualities and suppress them in themselves.

This leaves women as the one-half of humanity most likely to respond to beauty and to little children and young animals. We are the ones who can be emotional. We are the ones who bear the children and know of the effort it takes to raise a child into adulthood. We are the empathic gender and therefore the half of humanity who are more likely to hear the cries of the world, of Mother Nature, of Gaia.

The message from Mother is urgent. The half of humanity in charge of the world's agenda is led by men addicted to power and maintaining their dominance. Only now, there are weapons of mass destruction that can cause more suffering in a shorter time than ever was even imaginable. And, if patriarchal religions continue to exercise control over women, there will soon be more people than the Earth can sustain. Our beautiful blue and white planet, this garden island in space, our Mother the Earth needs our help.

It is time to gather the women to save the world.

Untapped Source of Peace

Untapped source of peace,
The only real hope
Is to draw upon the collective wisdom of women.
Those with direct experience of the cost of war:
The life of child, grandchild, sibling, spouse,
The loss of limb or mind of someone near and dear,
The loss of laughter, the pervasiveness of fear,
The loss of hope for the future.

Untapped source of peace,
Those who have known domestic violence:
Seen the effect of bullying on sons,
Seen daughters become silent,
Seen light go out in their eyes.
Those who know
That when every child matters,
When none are hungry, abused, or discounted,
The world will become a kinder place
For us all.

Untapped source of peace,
Women with empathy
Who live in a world apart,
Are safe, loved, and fortunate,
Yet can imagine
Being helpless, beaten, and raped,

Then forced to bear a child
Conceived in violence.
Women who know in their hearts
That what happens to any woman
Anywhere
Could happen to them.

Untapped source of peace,
Women who see loved ones filled with vengeance and hate,
Hypervigilant, fear-ridden, or afraid to sleep
Because of the nightmares.
Husbands, brothers, sons, and now daughters
Home from wars,
Bearing little resemblance to who they could have been
In a peaceful world.

Untapped source of peace,
Women in circles,
Women connecting,
Women together
Bringing the sacred feminine,
Maternal instinct, sister archetype,
Mother power
Into the world.

Jean Shinoda Bolen

3

MONOTHEISM/ DOING WITHOUT MOTHER

The three monotheistic religions—Christianity, Judaism, and Islam—have several similar characteristics. All three go back to the same forefather, Abraham. All have holy scriptures, a written Word of God. All three are hierarchical and patriarchal institutions. All sanction war—against each other.

Religion is used to support and justify war. Both sides pray and have similar basic beliefs: My God is better, stronger, more moral than your God. Divine revelation is on my side: my interpretation of the word of God and the mind of God is right, you are wrong. Divine right is personally claimed: God has chosen me and given me power over others. God is on my side; my enemies are God's enemies. Winning is evidence of God's favoritism. Winning is a triumph, losers are humiliated.

Women and children always suffer and have had no say in the matter of war. That women should stay in the kitchen, mind the children, and leave the running of the world up to men is the message that women have heeded. The urgent message from Mother Earth and the archetypal feminine is to gather the women and save the world, because leaving it up to men is going badly, and will be catastrophic. The premise on which men rule the world is religion, with monotheism the form of religion that is bringing us to the brink.

Before women can show up, speak out, and work together, we need to be spiritually empowered to bring the human family together into a planetary community. Toward this end, women need to bring intuition and intellect and their own spiritual experience to bear upon religiously based assumptions. There are lessons to be learned and empowerment to be gained by women through sharing religious information and personal spiritual experiences.

The two common sayings that may be helpful when religions are examined from the standpoint of women's experience and intuition are, "The emperor has no clothes!" and "Don't throw the baby out with the bath water!"

Religion and War

Patriarchal history has been about power, extension of power, possessions, and control. Western civilization began in ancient Greece. Athens was the first center of power and culture, later superseded by Rome. Both ancient cultures had a similar pantheon of gods and goddesses, were patriarchal in structure, and established supremacy through military power and war. Wars extended boundaries, established control over conquered lands, and took possession of sources of wealth, including slaves. Only after the Roman Emperor Constantine's conversion to Christianity in the fourth century did religion become another reason for the justification for war. Since then, there have been wars of religion, wars over religion, wars to convert pagans, to suppress heresy, and, of course, the Crusades to fight infidels.

Fratricidal Wars

Civil wars and wars between neighboring people are fratricidal. So are the current hostilities between adherents of the three monotheistic religions. Judaism, Christianity, and Islam all trace their origins to Abraham, who was the patriarch and progenitor of each of these religions. His descendants through Isaac, the son of his wife Sara, and his grandson Jacob established Judaism. Jesus and Christianity came out of this lineage. The prophet Mohammed, the founder of Islam and the Muslim religion, was the direct descendant of Ishmael, another of Abraham's sons, born to the slave woman Hagar. Descendants of these half-siblings have been warring against each other, with the Middle East the battleground. From the Crusades, which pitted Christian against Muslim, to the conflict between Israelis and Palestinians, to the current war in Iraq, religion plays a huge part in the ongoing conflicts.

Wars of religion can be seen as expressions of the extreme hostility between brothers who are jealous of each other, or want revenge for prior humiliations, or are exercising the power they have because they are stronger and they can. War is fratricide or sibling rivalry at its worst, and has as a model the story of Cain and Abel—the very first two brothers in the creation story, told in Genesis. As many people know, Cain killed Abel.

As a psychiatrist, I'm interested in the psychological provocation: Why did Cain do this? There is an explanation in the text. Abel was a keeper of sheep, and Cain was a tiller of the ground. Each brother brought the best that they had produced to the Lord (God), who had high regard for Abel and his offering, and no regard for Cain and his offering. When this happened, "Cain's face fell and he became angry." Later, when the two of them went out into a field, Cain killed his brother. In a family, when one child is arbitrarily favored and praised, the favored child reaps the jealousy, envy, resentment, and hostility from the disregarded, rejected child. Cain was responsible for what he did, but the situation was created by the God of the Old Testament.

The genocide in Rwanda at the end of the twentieth century had very similar fratricidal dynamics. Two tribal peoples—the Tutsi, who were mostly cattle owners (similar to Abel), and the Hutu, who were mostly farmers (like Cain)—had lived side by side, probably for generations and generations. When European colonial powers came to exploit the resources of Africa, and rule, Belgium colonized the land that is now Rwanda. Under the hierarchy established by the colonialists, the Tutsi minority were favored and elevated, while the Hutu's majority were arbitrarily refused opportunities for advancement. This caused deep resentment among the Hutu toward the favored Tutsi and their higher status and material possessions. When Rwanda became independent, the Tutsi were the privileged and educated managerial class. Hate radio fanned the long-standing resentment of Hutu men, and through demagoguery and the distribution of machetes, genocide was instigated, which resulted in 800,000 deaths in one hundred days.

Obedience to Authority: Abraham's Willingness to Sacrifice Isaac

Obedience is a religious virtue as exemplified by Abraham, who proved himself to God by his willingness to kill his young son Isaac. In the biblical story, Abraham invites Isaac to travel with him to Mount Moriah, which is several days away. Once there, Isaac eagerly helps his father make an altar of stones and gather firewood for it. It is easy to imagine young Isaac's excitement and participation in this important task. Then, in innocence, he asks, "Where is the lamb for the burnt offering?" Imagine the shock, betrayal, and fear that Isaac must have felt when he realized that he was to be the offering. Abraham bound Isaac, laid him on the wood, took out his knife to slay him, and was about to do as he was ordered when, according to the biblical story, the angel of the Lord intervened. Abraham's obedience had been tested; his willingness to kill his son pleased God.

Abraham is praised and blessed because he would kill his son if God ordered it. He is a religious example of unquestioning obedience, which is a military virtue. Soldiers follow orders that come down from above. Young men are willingly sacrificed by patriarchs when they are sent to war.

I doubt that Abraham told Sarah that he planned to kill Isaac. She probably waved good-bye and was without a clue that this was supposed to be the last time she would see her little boy alive. Any woman with a shred of maternal instinct would have at least protested. As Teri Wills Allison, a mother of a soldier and a protester of the war in Iraq, said, "I am not a pacifist. I am a mother. By nature, the two are incompatible, for even a cottontail rabbit will fight to protect her young."

Lost in Politics: A Loving God

When I hear spokesmen for Christianity, it makes me wonder what happened to the message of love that was the foundation of Jesus' teaching. When I was a child in Sunday School, I remember the words "God is Love" on a classroom wall. I sang "Jesus loves me, this I know, for the Bible tells me so." The Christian message I was taught was about love, forgiveness, mercy, comfort, and healing. The two great commandments that Jesus said summarized all the law and the prophets were to love God and to love your neighbor as yourself. He also said such words as "What you do to the least of them, you do to me," "Love your enemies and pray for those who persecute you," "Judge not that you not be judged," "Do to others, as you would have them do to you."

The Old Testament Hebrew prophet Micah said much the same: "And what does the Lord require of you but to do justice, and to love kindness, and to walk humbly with your God," which was the quote I chose for my father's headstone.

The prophet Mohammed had a great reputation for honesty, simplicity, and kindness. In Islamic teaching, Mohammed spent months meditating in a desert cave near Mecca, where the angel Gabriel appeared to him and he heard the voice of the Lord, which he wrote down. This became the text of the Koran.

There are many anecdotes about his compassion for the poor, of his helping women, his kindness to animals, and his pardon of enemies, even the man who killed and mutilated his beloved uncle. Stories about Mohammed and those told about Jesus in Matthew, Mark, and Luke could easily be about either man.

Instead of a focus on love, spokesmen for Christianity (the Religious Right and the Vatican) condemn and instill fear. They oppose women clergy, same-sex marriage, birth control, teaching evolution, feminism, and liberalism. Fundamentalist Christians look forward to Armageddon and the end of the world. Conflict in the Middle East is welcomed by some Protestants as a sign of the Second Coming. Meanwhile, Muslim fundamentalists talk about purification in the name of Allah and recruit terrorists.

Tenets of Fundamentalism

When *fundamentalist* is used as an adjective for Christian, Muslim, or Jew, it usually means the more extreme, politically right wing, literal interpreters of their respective scriptures and of doctrines justified by taking words out of context. Fundamentalism is also an evangelical Christian movement that arose in the United States in the second decade of the twentieth century, and is composed of different groups with similar beliefs, the largest of which is churches affiliated with the Southern Baptist Convention. Significant similarities among Muslim, Jewish, and Christian fundamentalists include beliefs that *their* scriptures are the Word of God and are to be taken literally, that doctrines derived from scripture are more important than love or compassion, and that men are superior to women. The fundamentalist believes that he is right and knows what is best for everyone else. He believes he has a right and responsibility to impose his point of view on others, especially women.

Religion becomes a simple matter when blindly followed, as stated on a fundamentalist bumper sticker (not a joke): "God said it. I believe it. That settles it."

To die on the battlefield and go to Valhalla was an ideal death in the dim pagan past. This is echoed in promises of a similar afterlife by Muslim fundamentalists recruiting suicide bombers. Numerous virgins are part of the afterlife paradise in both. Most fundamentalists—men and women—are male chauvinists, and their doctrines have very little to say about the afterlife for women.

Father Rules

The patriarchal family model is based upon father rule: sayings such as "A man rules the roost," "A man's home is his castle," "His word is law," and he must "wear the pants in the family" are colloquial expressions of this authoritarian model. Since the opposite of patriarchy is equality, feminism is a threat to this "natural order," and egalitarian marriage is a direct threat. If one person's ambition, feelings, needs, and perceptions must be followed, everyone else is subordinate and less important. Dys-functional family psychology is based on this same model—

one person is the narcissist, the other, the codependent. When the father of a family or the father of a country is a dictator, the flaw in the model is most evident and suffering is the cost to others. Women can behave in this same dictatorial fashion in some families, but that is not the general rule, and not the model supported by the culture.

In a healthy family, everybody matters, everybody's needs are considered, decisions are made together, and everyone takes responsibility for his or her part. People are safe and look after each other. Leadership is taken and shifts among those who can do what is called for and who has the time and is willing. When conflicts arise, resolution comes through communication, sometimes with the help of prayer. Love, rather than power, is the operating energy in a healthy family. There is laughter and spontaneity.

Monotheism is used to support authoritarian male leadership. For women, what is subtly or specifically emphasized in monotheism is that the male God favors men and gives men dominion over women and children and all living things. The more authoritarian the religious leader, the more patriarchal in attitude and beliefs that religion or congregation usually holds about the place of women. This is especially justified by theology that literally sees men as made in the image of God. Their hierarchical position is then given by divine right. Like the effects of discrimination on the basis of race or sexual orientation, the effect on women who accept this is an internalized lowered self-esteem, sometimes bordering on shame for belonging to an inferior gender.

God, the Authoritarian Father

In my consultation room, I hear a lot about fathers. The authoritarian father who can be terrifying to his children when he becomes angry and loses control is one all too unfortunately common example. The God of the Old Testament models this behavior. The biblical story of the Great Flood is a case in point. As the story is told in Genesis 5, God created man in the likeness of himself, creating both male and female at the same time (in the more familiar Genesis 2 version, God made Eve from Adam's rib) and blessed them. One chapter later, the Lord was sorry he had created humans because they were wicked, and decided he would blot out humans, beasts, creeping things, and birds. "Everything that is on earth shall die." He saved only Noah and his family, and the ark filled with pairs of animals and birds. After the flood, God promised he would never again destroy every living creature as he had done, and he placed a bow in the clouds to remind himself and humanity of this covenant. The Lord's wrath was next directed toward the cities of Sodom and Gomorrah, upon which he rained down fire and brimstone. He destroyed the cities and the valley, all the inhabitants, and everything that grew on the ground.

Television coverage of the results of the 2004 tsunami in Southeast Asia and of floods elsewhere give us some idea of what it would have been like to have water rise up until it covered the Earth. The story of the Great Flood was also an example of collateral damage on a planetary scale. While less damage was done when he rained down fire and brimstone, everything was destroyed, innocent and evil alike. And once more, the suffering of women, children, and animals was of no consequence. They suffered the fate of the men the Lord set out to punish. Destruction on a similar scale could happen again, this time brought about by men acting like the Old Testament God. All it would take is an authoritarian leader of a country with nuclear missiles to be angry enough and short-sighted enough to begin a nuclear war.

Through a psychological lens, the image of the Lord is of a controlling father who overreacts and justifies what he did by blaming the children for bringing it on themselves. Living with an authoritarian father, children learn never to question anything and to accept his rules and punishment.

In the New Testament, God wants to reconcile himself with humanity. To forgive his creation, from whom he is estranged, he requires that his son Jesus be crucified. Unlike Isaac, who had no say in the matter, Jesus is an adult who accepts this task. The more fundamental the Christian, the more the focus is on the suffering figure of Jesus. The message is, "He died for your sins."

Every mother can imagine how awful it was for Jesus' mother Mary to be at the foot of the cross through those long hours, witnessing the suffering of her son, and hearing him cry out in despair, at one point, "My God, my God, why have you forsaken me?" It is no wonder that Roman Catholics and Orthodox Christians turn to Mary and pray to her for mercy. Especially mothers.

Mysticism and Fundamentalism

To be "born again" is a profound, mystical experience of being loved just as you are by a loving and forgiving God—especially so in Billy Graham's ministry. The message, music, and charismatic ministry make it possible to open the heart and feel that you matter and have a place in a benevolent universe. In evangelical Christianity, the peace that comes through the experience is attributed to Jesus Christ, the Holy Spirit, and God the Father. It is a numinous experience of oneness with divinity that is wordless and ineffable, a non-ordinary reality.

Back in ordinary reality, the born-again experience becomes attached to the idea of eternal damnation or eternal reward, by a patriarchal religion that emphasizes fear and obedience. For many, it is difficult if not impossible to separate the personal transformative experience from religious dogma that binds them to also believe what they are told God has said, based upon the literal interpretation of selected parts of scripture.

While there is a diversity of religions and absolutism in every fundamentalist belief, spiritual and mystical experiences are universal. They occur not only in a religious context but often in the vastness and beauty of nature or in moments of crisis.

Christian women and men who think for themselves have to grapple with the discrepancy between a personal experience of a loving and forgiving divinity and religious teachings that condemn all those who haven't heard of and accepted Jesus as their personal savior. The list is long: everyone born before the Christian Era, or born in places where the message didn't reach, all those who practice compassion and yet are not Christian, such as the Buddhists, including the Dalai Lama.

There is also the disillusionment problem in all hierarchical religions that follow the dominator-patriarchal model, in which men must be looked up to, obeyed, and seen as exemplars of divinity. Sexual exploitation of vulnerable followers, such as pedophilia that was committed by priests and covered up by the Roman Catholic Church, revelations of sex between gurus and their devotees, and ministers caught with pornography or prostitutes are all common causes of disillusionment.

Religious laws, attitudes, and practices adhered to without compassion are another problem, especially for women, who also suffer the most from them. Women suffer directly from fundamentalist religions in a variety of ways. Raped women are victims who suffer further when they are blamed and punished for it, and if impregnated must bear the child. There is the cruel practice of female genital mutilation, done to ensure virginity, that is practiced by Muslim fundamentalists, mostly in African countries.

Spiritual Women and Patriarchal Religion

Staying within any religious tradition once the shadow effects of power are seen is a challenge. One possible exploration is to see the religion through a wide-angle lens rather than a narrow aperture. All three monotheistic religions have mystical traditions: the Kabbalah in Judaism, the Sufi in Islam, and Christian mysticism. All three also have liberal traditions that include treating women more as equals than is acceptable to their fundamentalist brethren. Within each religion, theologians have differences, and there is a vast literature about those differences. There is also the historical context, which when taken into consideration sees many practices in scripture as reflections of life at the time.

In church history, as well as history in general, the victor's version of history prevails. In Christianity, once the Roman Emperor Constantine made Christianity the official religion, the now politically empowered and misogynistic church fathers defined which Christians were heretics and what was orthodox and acceptable to believe. Among those condemned as heretics, whose beliefs and practices were once known only through their persecutors, were the Gnostic Christian churches, many of which had women leaders. These early Christians were egalitarian; women and men gave sermons and communion. All of their scriptures, some written at the same time as the earliest in the New Testament, were deliberately destroyed by the church fathers as heretical. Then, in the mid-twentieth century, copies of the Gnostic scriptures were found sealed in clay jars in a desert cave near the Egyptian town of Nag Hammadi. They included teachings of Jesus by disciples, the revelations of mystical writers, and a theology that was not identical to that which came down through the church fathers. A current reevaluation concerns the importance of Mary Magdalene. She may have been the most important of Jesus' disciples, though she was labeled by Pope Gregory the Great as a prostitute.

It is helpful for women to realize that religions are shaped by men whose egos, psychology, and motives would affect the doctrines. The task of growing spiritually is similar to making inner peace with a dysfunctional family. It is important to feel the pain of what was done in the name of religion, to realize that religious leaders (like parents) can do wrong, and that you (and others) did not deserve neglect or abuse. Compassion and forgiveness of yourself and of them is called for. To stay within a religious tradition once the flaws and evils done in its name are acknowledged also requires an ability to hold the opposites, seeing the light and the dark.

Circles with a Spiritual Center

Spiritual empowerment, like all other forms, comes through becoming aware of unexamined assumptions and personal experiences. A spiritual consciousness-raising circle can be a support group, a silent prayer group, a study and discussion group, or a sanctuary for all three. If each woman in the circle spoke from her own religious experience, and could voice her perceptions and questions, it would become a source of spiritual and psychological support. Once a circle becomes a safe place to tell the truth of one's own experience, a collective trust and wisdom emerges.

If women in churches, synagogues, mosques, and temples gathered together and met in circles with a spiritual center—meaning that divinity was invited into the center, as a silent presence—changes would slowly happen in the women and in their homes and churches. When women listen to each other's stories and share their own, growth happens and confidence grows. As harmless as such an endeavor may seem to women who like the idea, this may be a threat to the established order. Men in authority worry about women talking freely to each other. As well they might.

If women created circles where they worship, and became an influence within their religious communities, there would be truth to the bumper sticker that says, "It's my church, too."

Spirituality and Sexuality

Human beings are created with the capacity for spiritual epiphanies just as we are for sexual orgasms. Both are ecstatic experiences that are "built in at the factory" as Jungian analyst Joe Wheelwright often said of matters archetypal. Patriarchal religions and governments influenced by fundamentalists exercise control over women's sexuality. Fundamentalists maintain that a woman's sexuality is for reproduction or the pleasure of the man, making it difficult if not impossible for women to be orgasmic when sex is her marital duty, and enjoyment sinful and cause for guilt. Yet women are not only inherently orgasmic but can have multiple orgasms, which means that women have a greater capacity for sexual pleasure than men, who are not as physically able. The loss of control over women's sexuality threatens patriarchal authority. The loss of control over a particular woman threatens a man.

Likewise threatening are ecstatic mystical epiphanies, because they are uncontrolled by religious authority. During the Inquisition, an altered state of consciousness and witchcraft became synonymous. Women with ESP, mystical, or hallucinatory experiences were burned at the stake. Spirituality is often enhanced by trance states with or without hallucinatory drugs. The opposition to medical marijuana and the criminalization of all hallucinogenic drugs has parallels to the suppression of contraception: they make it harder to have ecstatic, sense-enhancing experiences. Altered states of consciousness are most accessible to those with access to the right cerebral cortex and can easily enter trance states—people who may be poets, artists, meditators or healers. Those who are more typically rational-minded, linear-time oriented, and goal-focused access the left-cerebral cortex and have a more difficult time entering trance states. Women are the more able gender in this respect, also.

Jesus, who spoke in parables, healed, was nonhierarchical, had visions, and was a threat to the established order, had a lot more in common with women than the men who speak in his name. From what I know about the prophet Mohammed, the same could be said of him.

The Missing Mother Goddess

The devaluation of the Sacred Feminine and women coincided with the rise and domination by warrior people whose chief or only divinity was a sky god who was intimidating, punishing, and powerful. The Greek god Zeus on Mount Olympus and his Roman counterpart Jupiter ruled. Each had thunderbolts that could strike from a distance and punish those who displeased them. Expecting to be "struck by lightning" is still an expression we use when we fear that we may have gone too far in our independence. The Judeo-Christian God was associated with Mount Sinai, where Moses received the Ten Commandments, was referred to as a heavenly father, and also was feared.

Under patriarchy, the Mother Goddess as an archetypal or divine maternal force was driven into the collective unconscious, where anything suppressed is forgotten by the ego but remains as a potential that can be remembered. It is the nature of the psyche that anything suppressed, denied, and cut off from consciousness can later be remembered and brought back into conscious life.

The Sacred Feminine is coming back, mostly through women, but also men who have dreams of numinous goddess figures or who follow intuition and instinct to find and resacralize land or go on pilgrimages. The success of fiction such as *The Mists of Avalon* in the 1980s, and currently, *The Secret Life of Bees, The Da Vinci Code,* and others that delve into the figure of Mary Magdalene are also indications that the missing feminine element holds deep meaning. In some Protestant churches, prayers are directed to "Mother-Father God"; in Judaism, more emphasis is given the *Shekinah,* a feminine divine presence; in the Orthodox church, an increased emphasis on Mary as *Theotokos,* a Greek title that has no English equivalent, although it means "the God Bearer." In Catholicism, the apparitions of Mary, the Mariology movement to elevate her further, and the increased importance of the Black Madonnas are all indications.

Whether referred to as Goddess, or called the feminine face of God, or by a particular name, the Sacred Feminine represents maternal concerns, the feminine principle, and compassion. Goddess is not "God in drag," which is an irreverent way of saying that this is much more than a female divinity with the same agenda and attributes of God the father.

Humans probably have always had mystical experiences and spiritual revelations. Paleolithic images painted on cave walls, some 40,000 years ago, suggest this possibility. Divinity is a great mystery, beyond comprehension by our intellect. The *Tao Te Ching* by Lao Tsu begins, "The Tao that can be told is not the eternal Tao. The name that can be named is not the eternal name. The nameless is the beginning of heaven and earth." The divine, here called the Tao, is beyond the grasp of the mind. However, as our ability to conceptualize and draw inferences grows, so might our concept of divinity. Monotheism as defined by patriarchy is a hierarchy with a male God at the top. Monotheism could also have an expanded meaning, as an eternal Oneness, or Unity, from which everything evolves. The mystical moment is an opening to this field or realm that feels divine, loving, beyond male and female, and yet both.

To hold a whole picture is to know that spirituality and mystical experiences are universal on the one hand, and on the other, that religions are a communal means through which people worship, pray, and develop rules of behavior. It is important for women whose religions define them as inferior to men, based on the gender of God, to learn that before there was God, there was the Goddess. In ancient Greece, the original trinity was the triple goddess as maiden, mother, and crone. Eleusis, just outside of Athens, was the site for the Eleusinian Mysteries, where for more than 2,000 years, until the temple was destroyed in 396 CE, people came and participated in a ritual and mystical experience, possibly with the aid of a hallucinogenic drink, and overcame the fear of death. It was Persephone, the divine daughter, rather than Jesus Christ, the divine son, who returned from the underworld realm of death, but the message, that death was not the end and need not be feared, was essentially the same.

Awe of the supernatural or divine is archetypal. There is in us all a tendency toward the spiritual—an orientation toward an invisible presence, to something greater than ourselves that cannot be fully known. Spirituality unites us—in silence, in awe, in devotion, and in soul connections. Patriarchal religions divide us into the saved and damned, heathen and Christian, believers and infidels, and other either/or categories. Women can move easily out of either/or dichotomies to be with women of other faiths. Seeking to find ways to feel related, we look for what makes us similar rather than what separates us.

In Jerusalem, for example, the Women's Interfaith Encounter Association brings Muslim, Christian, Druze, and Jewish women together. They are a Cooperation Circle of the United Religious Initiative. Motherhood unites them in this common cause: "We women of faith, as the creators and bearers of life, are saddened and upset by the continual killing of our children in the war and violence of the Holy Land." Meeting in circles of friendship, they listen deeply and learn "from the wisdom of the three religions that nourish our region."

If there is to be Peace on Earth, it will be because "Mother" returned home through her daughters.

4

MOTHER NEEDS *You*!

The Message from Mother to all the women of the world: Wake up! Arise! Do not ask for permission to gather the women. What cannot be done by men, or by individual women, can be done by women together. Earth is Home. In all the galaxy, there is no place like this home. It is time to take loving care of planet Earth and all life upon it, beginning with where you are, what you love, and your vision. The three chapters that follow this one lay out how we can do this. This chapter is like a mirror for women to see themselves: Look! Women have qualities that men have not developed, and these talents are needed right now.

If Mother Archetype, Mother Goddess, Mother Earth, the deep feminine placed a classified ad in the "Help Wanted" section, the attributes needed would apply to most women and some men. Individuals do matter, but in the matter of necessity—to save the world—the need is for the women of the world who make up more than half of the population to take on the job. The ad might read:

> HELP WANTED: Everywoman. Home keepers for Earth. Must keep premises safe for all. Have concern for children's needs and development, ability to manage resources, resolve conflicts, work collaboratively, ask questions, listen, and learn from the experience of others, be empathic, and act with compassion for the benefit of all, including generations to come.

The qualities that I focus on as characteristic of women are generalizations that hold true in a patriarchal world where gender differences are emphasized, and boys and men are afraid to be like women. How much is innate versus a result of socialization will not be known until the dominator-patriarchal paradigm of power over others evolves into something new. Paul H. Ray and Sherry Ruth Anderson described the subculture of *The Cultural Creatives,* some 50 million women *and* men in the United States who subscribe to the values in Mother Earth's classified ad. The men in this category are archetypal brothers rather than patriarchs.

The distinguishing female abilities that contrast with standard male behavior have to do with qualitative differences in response to stress and uses of communication. These differences relate to women's ability to manage resources responsibly, to work together for the common good, and to base ethical behavior upon compassion for others.

I also think that it is important for women to understand how identifying with the aggressor and going along to get along are major lessons learned in boyhood and young manhood. This is why I devote parts of this chapter to these subjects. Mothers who remember a little boy full of questions, wonder, vulnerability, and feelings may understand their sons and patriarchy better as a result, and also see why Mother needs us.

Women's Tend and Befriend *Response to Stress*

In 2000, women researchers at UCLA noted that they reacted differently from their male counterparts to a stressful work situation, and because they were scientists took the next step and researched this difference. The women came in to talk about what was going on and how they were affected, cleaned the lab, had coffee, and tended to whatever needed doing. Meanwhile, the men in their department retreated into their offices or laboratories and isolated themselves, which is behavior consistent with the flight aspect of the "flight or fight" reaction. That women behave differently than men had not been noticed before, and the male reaction was considered the normal human response. This was typical of research that used male subjects and then applied findings to men and women, a standard procedure before women became researchers.

Shelley E. Taylor, the principal investigator, and the team of women researchers found that females of many species, including humans, responded to stressful conditions with a *tend and befriend* response. Females protected and nurtured their young, and sought social contact and support from others, especially other females.

When women's lives become stressful, they talk to their friends, look after their children and animals, straighten up the house, water their plants, clean the lab, studio, or desk where they work. And as they do, their stress level goes down, and the amount of the maternal bonding hormone, oxytocin, goes up.

If something startling or unexpected happens in the midst of other people who are strangers, women reduce their stress through social contact with them. They turn to whomever is there and engage them in conversation. If a woman is driving and gets lost, this is a stress that she reduces by stopping to ask directions from a stranger.

Men's Flight or Fight *Response to Stress*

Men, on the other hand, had an increase in adrenaline when they were stressed, a reaction that prepares their bodies to run or to stay and fight. It is a physiological reaction that is enhanced by testosterone, while the "tend and befriend" response is enhanced by estrogen. Oxytocin is called the maternal bonding hormone because it was first studied in childbirth. It is a hormone secreted in both men and women, just as adrenaline is. Oxytocin buffers adrenaline and has a calming and relaxing effect.

Men also reduce their stress and benefit from the calming effect of oxytocin in safe conversations, where feelings and vulnerabilities can be talked about without being put down or feeling one-down. This usually means in conversations with women or in therapy or recovery groups such as Alcoholics Anonymous or Al-Anon. In my office, I can see how much talk helps men as well as women reduce anxiety, anger, and isolation, which is not only calming, but centering.

Work is especially stressful wherever people are treated as expendable cogs in a company machine and are casualties of mergers, downsizing, or exportation of jobs. In nonprofit situations, if funding sources dry up, it doesn't matter how important the work is. Men more than women have their worth defined by work because they are much less invested in their relationships or find gratification in them. When job stress or unemployment becomes a possibility, which is a sign of the times in the United States, the male half of the population is afflicted by the flight-or-fight adrenaline-testosterone response. The result is an increase in rage and in withdrawal. Men become diffusely angry, which comes out as road rage, or as domestic abuse, or as scapegoating others. Or, there is withdrawal into leave-me-alone behavior, with men retreating into the virtual world of television, the Internet, or electronic games.

Where there is chronic unemployment and little or no hope for work, men's rage covers their feelings of impotency and lack of respect for themselves and by others. In everyday life, some men compensate for psychological impotency by exercising physical dominance over their families and women with threats of violence. The plague of violence in the home, in street gangs, and by terrorists is a reflection of an economic and social world where Mother as the feminine principle is absent, and no one cares about each person in the human family.

Women Talk, Learn, and Bond

In villages and neighborhoods, mothers gather and talk, share information, gossip, laugh together, keep an eye on the children, and do whatever task brings them together. In Third World countries, they may be washing clothes at the edge of the river or filling water containers at a pump or well. In office buildings, women talk personally on coffee breaks. In another era, they would be talking at quilting bees. On college campuses, women organize meetings, talk into the night, form lifelong friendships. Old friends get together for reunions and stay current with each other through decades of change. Most women value and sustain relationships with each other.

Women bond through conversations in which rapport and trust grows through what we tell each other and how we respond. We talk about ourselves and about our relationships. We keep current about matters raised in previous talk, which may be about troubles. Friendship is a matter of depth, of mutual self-revelation, of being able to be uncensored, unwary, and vulnerable. Women's talk may also lead to taking care of one another's children or animals, or trading a work day, or accompanying one another to the doctor or to the hospital. We also keep up with each other through what we hear from mutual friends.

I need to add here that how I am characterizing women's talk is another generally true but not always so statement. Women's friendships are expressions of the sister archetype, while the fluidity of nurturing and sustaining that goes back and forth depends upon women also functioning from the mother archetype. These archetypes are not the important ones in every woman.

Women's talk is called gossip when the subject is someone else. Yet the spirit in which the conversation is held makes all the difference. Gossip turns malicious when colored by envy or feelings of superiority. This is gossip that deserves the bad name. It creates an atmosphere of mistrust and unease and is about hierarchy and status rather than the usual function of women's talk, which is to promote closeness as equals, to keep in touch with how people we know are doing, and the learning that comes when insights are part of the conversation. Good gossip is motivated by love and concern; it keeps mutual friends who care updated about each other.

Women learn through conversation. Women's stories can inspire "If she can do that, I can, too!" responses. They also provide warnings to be wary; a woman listens and takes to heart "What happened to her could happen to me." Women have always exchanged practical information as well, from recipes to professional referrals. There is empathic support, problem solving, and stress relief in women's conversations. It is what we as a gender have a natural talent for that has been encouraged by our social roles, especially as mothers. Women's talk leads to symmetrical connections, rather than hierarchical ones, and consensus building, both of which are needed for cooperation and collaboration.

Women's talk needs to be recognized as a positive force—a means and method of bonding and understanding each other and other people—that humanity needs to be able to take a next step toward planetary community.

Deborah Tannen brought gender-based communication styles into public consciousness in the 1990s with her book, *You Just Don't Understand.* Based on research, of others and her own, she wrote about gender differences in how men and women speak. She found that a woman talks "as an individual in a network of connections. In this world, conversations are negotiations for closeness in which people try to seek and give confirmation and support, and to reach consensus. They try to protect themselves from others' attempts to push them away."

Male Hierarchy and Men's Talk

In contrast, Tannen emphasized that a man talks as "an individual in a hierarchical social order in which he is either one-up or one-down. In this world, conversations are negotiations in which people try to achieve and maintain the upper hand if they can, and protect themselves from others' attempts to put them down and push them around."

In fieldwork studies of animals, males who dominate others are called "alpha males." This expression is now often used to describe a man who is and also behaves like a top dog. Conversation among men serves the purpose of establishing who is alpha to whom, and in a group, who has the most status. When men ask, "How's business?" or talk about their cars, or sports statistics or stocks, or mention what they do and where they work, schools and clubs, it is information that establishes ranking. Men get along with each other by finding their place in the pack or knowing their place. In a group conversation, the alpha male either dominates the conversation or the talk is subtly directed toward him, or he is looked to for the last word.

Alpha males challenge each other and establish dominance through talk. When alpha males are involved in adversarial negotiations, their personal agendas are likely to trump the best interests of the people they represent. Alpha males may be forced by politics to take part in a peace process in which they face each other across a table, but since peace accords require the ability to compromise, agreement is unlikely if each man defines this as a concession that puts him one down to the man he desires to humiliate.

Humiliating the enemy is a motivation that perpetuates conflict into the next generation. Ariel Sharon in effect put Yasser Arafat under house arrest in his headquarters in Ramallah by surrounding the building with Israeli tanks and troops. This was humiliating enough, but to also cut off utilities so that there would be no air conditioning or plumbing added further insult and real discomfort. Arafat's status was reduced and his humiliation was shared by males in the Arab world who identified with him. It can also be surmised that testosterone levels were affected. Joshua Goldstein in *War and Gender* notes that it is not just the perception of status that fluctuates but also the level of testosterone. In the male competitive world, testosterone levels begin to rise in anticipation of the competition; then they continue to go up in the winners, but fall in the losers.

In a women's workshop at the United Nations, I heard about the peace process in Sierra Leone, which finally was settled in 1999 after women were included in the talks. Their presence was initially objected to by one of the male leaders who complained contemptuously about the presence of women, saying, "We don't want the women, they would just make compromises!" We all laughed. It was, after all, precisely why women need to be present.

Peace negotiations are stressful: it would be beneficial to have more oxytocin and estrogen to dilute the adrenaline and testosterone in the room. Or, in the case of the peace process that led to the end of "the troubles" in Northern Ireland, to have women go-betweens, when representatives of the Protestants and Catholics were initially not even willing to sit down together in the same room. Four Northern Irish women who did not take sides were involved in the peace process. They could go between the rooms and carry messages back and forth, while former U. S. Senator George Mitchell could hold discussions with each group. A peace plan was agreed upon and signed on April 10, 1998, which is known as the Good Friday or Belfast Agreement. This agreement provided for a transition period to disarm and form an interim representative government.

The women who were involved in the successful peace negotiations were members of the Northern Ireland Women's Coalition. The NIWC was founded in 1996 to work for "reconciliation through dialogue" and to increase the role of women in politics. It is nonreligious, against terrorism, and did not take sides. They had two seats in the interim assembly and have two seats in the new Northern Ireland Assembly. They saved the peace plan when a vote of no confidence that would have dissolved the interim government was defeated by their votes.

Peace accords that finally ended the conflict were worked out once women became involved. Terrorist acts of aggression and retaliation in Belfast and other cities had been carried out for over thirty years from 1968 to 2001. If peace could come to Northern Ireland after decades of fratricidal conflict, peace in the Holy Land between Israelis and Palestinians might also be possible with women as peacemakers.

Why Men Don't Ask Directions

Women laugh about how men would rather drive around for hours than ask someone for directions. It is a small inconvenience when a man doesn't ask directions and takes longer getting somewhere, but as an example of avoidance behavior and mistrust, it can also be appreciated as a tip of a patriarchal iceberg. It reflects experience that has become ingrained.

It is important to understand why this is so typically male, and what it says about men that something as straightforward as asking directions is actually an emotionally loaded situation going back to childhood and reinforced thereafter. Hierarchy is part of the playground and after-school culture of little boys. Men learn in childhood how painful it is to be made fun of when they don't know something that the other boys do. They learn that asking a question can reveal their ignorance *and* innocence, and are shamed. They also learn that when they ask a question, they can be deliberately misled. The other male might not know the answer and yet acts as if he does. He makes up an answer rather than reveal his ignorance. Or he may have fun at the younger boy's expense and give him a wrong answer on purpose. To be upset and angry at learning he was misled then sets a boy up for further humiliation. He is likely to be teased for his anger or made fun of for being gullible.

Playground Lessons

Boys can be goaded into fighting when unequally matched. It is a double-bind situation when he is picked on and goaded. If he doesn't put up a fight when taunted by another boy, he is humiliated further; if he fights, he'll be beaten. When a boy is in a physical fight, there are usually onlookers who shun him if he loses and cries. He is shamed. Fights can be avoided by boys who learn to use talk to diffuse tension. The fine art of negotiation is most developed by those who need to deflect hostility rather than by alpha male boys or men.

Boys learn that the male world is made up of winners and losers, or predators and prey, on the playground and streets. A boy with imagination, access to feelings, and intuition often understands that he cannot afford to express sympathy for a kid who is picked on, because the two of them would then be lumped together, and he would be excluded from the group or gang and be in a vulnerable outcast position himself. Boys who have been treated sadistically and humiliated unconsciously "identify with the aggressor." They do unto others as was done to them. A humiliated boy will take his pent-up, impotent rage out on others; or he plots and plans getting even, which helps him to feel powerful; or he turns it against himself, becomes depressed and potentially suicidal. Guns and explosives become an equalizer in fantasy or reality. Little boys with access to handguns have taken them to school, and a couple of them have used them on classmates. Older boys who killed others and themselves at Columbine High School had been picked on by the jocks in the school. It is likely a dynamic in teenagers who volunteer to be suicide bombers. Hierarchy and humiliation perpetuates patterns of aggression and retaliation.

If mothers in every school acted together to prevent the bullying of anyone, and saw the bully as a troubled child or adolescent who is in need of help, it would make the school a safer place, where learning and respect for others could take place. In contrast, an individual mother who stands up for her picked-upon son can make matters worse for him.

Abu Ghraib Prison Abuse

The photographs from Abu Ghraib prison were staged to mock and humiliate Muslim prisoners: they were naked, were forced to assume sexual postures, some were blindfolded and terrorized by American soldiers who were apparently having a good time. When the photographs were published and the world was outraged, radio call-in host Rush Limbaugh responded to a listener who said, "It was like a college fraternity prank to stack up naked men. . . ."

Limbaugh agreed, "Exactly! Exactly my point! This was no different than what happens at the Skull and Bones initiation and we're gonna ruin people's lives over it, and we're gonna hamper our military effort, and then we're gonna really hammer 'em 'cause they had a good time. You know, people are being fired at every day. . . . You ever heard of emotional release . . . of blowing some steam off?"

Male athletes and men in fraternities have an elite status on college and university campuses in the United States. From what has been revealed—often in investigations into the deaths of pledges during Hell Week in fraternities—Limbaugh and his caller are right. Team initiations in certain sports are also similar. Initiates are not prisoners, but they are treated very similarly to abused prisoners. They assume the position—which involves bending over, sometimes buttocks exposed—to be hit with wooden paddles. They are sleep deprived, humiliated in all manner of ways, eat disgusting things, drink to excess so that most deaths are from alcohol poisoning, may be kept naked and have sexual acts performed on them or by them. Very Abu Ghraib. Sadistic initiations are the price to pay to belong, and having survived the ordeal, they in turn will be the ones to inflict similar atrocities at the next initiation. This is an institutionalized form of identifying with the aggressor.

Women Manage Resources Responsibly

When mothers have money to spend, especially if there is little of it, they usually spend it on others and on necessities. Men spend money differently than women would. It is a male thing to have the most impressive weapon or vehicle. Other men are the audience that matters: the males he wants to impress or continue to dominate. On the world stage, we see leaders of their countries acting in ways that look like the behavior of men in Third World villages and in exclusive urban men's clubs. Getting respect from one's peers or gaining entry to The Club, whatever it may be, is the metaphor.

The most exclusive club in the world that a leader of a have-not country can aspire to is the Nuclear Club. Making and testing at least one nuclear bomb takes money that could be spent at home, improving conditions that would benefit the people. Similar choices are made by leaders of the United States, in amassing a huge debt to pay for a war and for all the shiny equivalents of aggressive boy-toys on a massive level. There is money to pay for weapons and vehicles and new and fascinating technology, but not for adequate medical care or medicine for anyone who needs it. Budgets for vulnerable or disadvantaged people and poor children are usually the last funded and first to be cut when men are in charge.

Budgets for the Human Family

The purse strings of the planet are held by men. The greatest expenditure: global military spending at $900 billion. In 2003, according to the Women's Environmental and Development Organization, the estimated funds needed to look after basic human needs were as follows: to provide shelter, $21 billion; to eliminate starvation and malnutrition, $19 billion; to provide clean safe water, $10 billion; to eliminate nuclear weapons, $7 billion; to eliminate landmines, $4 billion; to eliminate illiteracy, $5 billion; to provide refugee relief, $5 billion; to stabilize population, $10.5 billion; to prevent soil erosion, $24 billion. The estimated annual total budget for human needs, $105.5 billion vs. the actual global military spending, $900 billion. Imagine how differently women with maternal concern might manage the "family budget" now spent by the nations of the world.

Lessons from Microcredit Loans

In many parts of the world, with as little capital as $50 to $100, people can start a business. The institutions that grant microcredit to impoverished Third World people learned that when these loans went to men, the men usually spent the money on themselves and did not pay the loan back. Now most of these loans are to women who use the money to create a source of income and repay the loans. A poor woman might purchase laying chickens, a sewing machine, supplies necessary to do piece work out of the home, or capital to begin providing something that others will pay for. Across most cultures, women consistently deal more responsibly with loans than men. Microloans were started in 1976 by Muhammed Yunus, an economics professor, who created the Grameen Bank for this purpose in Bangladesh. There are now hundreds of institutions that grant small loans. Several million borrowers have received and repaid more than $2 billion in loans to the Grameen Bank alone.

The best credit reference for a poor woman is having a circle of women friends with whom she has discussed her plans for the loan and who will continue to be involved and supportive. Though loans are made to an individual woman, in the Grameen model, she must bring four others together into a group, which is the basic unit of its lending program. Two in the group can get the first loans; the other three are told that the borrowers need their support to be successful, and they are to help them use the money well and start earning. The loans are for a year, paid back in weekly installments. After six weeks, if conditions are going well and the loans are being repaid, two others in the group can get microloans, with the same requirement of support from the group. After another six weeks, if the borrowers are all paying their loans back, the last person can get her loan, again with the expectation of advice and support from the others.

Microcredit is a grassroots women's empowerment movement that builds community and draws upon the power of a women's circle to make it possible for them to do together that which they could not do alone. The women would reduce stress by talking about concerns that arise in their new businesses and about the reactions of husband, family, and neighbors. They would bring up ideas and suggest solutions, drawing upon their collective experience and wisdom. What one woman has learned, the others can use; what another has overcome is an inspiration to the rest. Borrowers also promise to fulfill the Grameen model's social contract, which is easier done with others than alone. Each woman promises that she will drink only safe water, limit the size of her family, educate her children, grow vegetables. In Bangladesh, an additional pledge is that she will refuse to participate in dowry customs.

These microcredit circles of women are contributing from the ground up toward the metaphoric millionth circle.

Water Rationing at the Faucet

The scarce commodity upon which all life depends is water. Through waste, pollution, and the low priority of providing clean water to poor people, in many impoverished parts of the world contaminated water is the only water available. Children are especially vulnerable to getting diseases from contaminated water. They die unnecessarily from preventable diseases and from diarrhea and dehydration that could be easily treated with IV fluids to make up for the fluid and electrolyte loss. The World Health Organization reports that 80 percent of all sickness in the world is attributable to unsafe water and sanitation. Water-borne diseases kill 3.4 million people, mostly children, annually.

Sometimes, in Third World cities or in villages, water is available only at certain times, through one main faucet, and then for only an hour or two a day. When this has to serve many individuals and families, the obvious problem is how to apportion it. If this were decided using the dominator model, the strongest male would appropriate this for his and his household's use, fending off challenges from others, leaving still weaker others to get whatever was left or do without. I wonder if men could do what I learned women did in this situation in India. They discussed the daily needs and numbers of people in each household, and once these needs were determined, the method was to divide the two hours that water was available into segments of time that each household could in turn use to fill buckets and containers with the precious water. For example, a couple might get five minutes; a family of six, fifteen minutes; and so on.

Water rights and clean water are emerging as crucial issues worldwide. Ownership of water determines whose land will flourish and whose children will thrive. Negotiations for water, like peace accords, need to involve women with maternal concern for life when decisions are made. Mother Earth and Mother Nature provide water in abundance; if it is up to mothers, clean water will be a right.

Women and the Ethics of Relationship

Before women entered the conversation about human values and gender differences, the authority of men and the natural order of hierarchy went unquestioned. Men saw women as lesser rather than different and equal, and male values were the gold standard against which any other alternative was lesser. One premise that had academic authority behind it was that men were the more ethical gender. Carol Gilligan's research undid this male model in which logic determined ethics. She established that ethics is not just about lower and higher principles, but also about compassion and concern for people. Principle and relationship: neither one, alone, is sufficient.

Carol Gilligan's *In a Different Voice* was published in the early 1980s. Through her Harvard research and the accumulated observations of many others, mostly women, Gilligan described an ethic of care or concern for others as an unseen and unappreciated ethical dimension. It therefore went unmeasured. This ethical dimension is much stronger in women than men. The Kohlberg test, which was supposed to measure ethical development, presented subjects with dilemmas. An example of one, and the responses to it given by two eleven-year-olds, beautifully demonstrates this significant difference between the genders.

This was the story of Heinz, whose wife is dying. The pharmacy has the medicine that will cure her, but he is too poor to buy it. He is in a predicament. The question posed on the test: Should he steal it?

The question assumed that there were only two choices; Heinz could steal the drug and save his wife's life, or he could obey the law and his wife would die. The test was constructed as a series of questions designed to reveal the underlying structure of moral thought. In it there is a hierarchy of values that could be logically arrived at, with the principle of justice the highest stage. The two eleven-year-olds, Jake and Amy, are tested. Jake accepts the assumptions of the test as he struggles with the dilemma and his reasons for thinking as he does, such as life has more value than property, or that the pharmacist's money is replaceable but Heinz' wife is not. Jake scores well for his age.

Amy, on the other hand, has trouble with the test from the beginning because she sees more options than she is given. In addition, she constructs the situation differently than the test does. Amy begins with the premise that Heinz shouldn't steal the drug, and his wife shouldn't die. She suggests other ways of getting the drug besides stealing it, like borrowing the money or negotiating with the pharmacist. She is concerned that if Heinz steals the drug and goes to jail, his wife could get sick again and he couldn't help her. She also suggested that Heinz and his wife should talk it out and find some other way to get the money. Since the test was designed as a series of questions meant to elicit logical answers, which didn't fit Amy's premise, she did poorly on it.

Amy, age eleven, on being presented with the problem of Heinz began by expanding the question from the either/or premise that Jake accepted without question. She thought about everyone's needs in this equation, which not only included Heinz and the pharmacist, but also Heinz' wife and the possibility that this would not be a one-shot solution.

In the real world, taking all this into consideration is also true. For example, when the problem is violence, solutions have to be found to keep noncombatants safe from physical harm and emotional trauma while taking the ego sensitivity of male leaders into consideration. From gangs in the 'hood, to Bosnia, Somalia, or the Israeli-Palestinian conflict, to whatever will come next in the Middle East that could involve nuclear weaponry, the gender best equipped to be empathic to everybody's needs and to anticipate how a current solution will create problems in the future is women with people skills in addition to intellect and knowledge.

Women Whistleblowers

A combination of idealism, altruism, and concern that what is going on is hurting others prompts women who do not want to make a fuss or be in the limelight to sometimes become reluctant whistleblowers. It is a step-by-step process to get there. Once the wrongdoing and the harm become clear to her, the next step is to let superiors, who could do something about it, know. When these steps are futile and feelings of moral responsibility do not go away, she then may become a whistleblower.

The desire to expose wrongdoing can also be motivated by revenge or reward (reporting tax evasion, for example, is rewarded by a percentage of the money collected) or on principle. However, I think that empathy and idealism, which has a component of innocence, are more likely to motivate women than men.

In 2002, instead of selecting a Person of the Year (formerly, Man of the Year), *Time* magazine named three women as Persons of the Year 2002. The three were photographed together on the cover, each with her arms crossed against her chest (a "Mother means business" pose), with the heading "The Whistleblowers" announcing why they were selected. Each had taken huge risks and been through ordeals to blow the whistle on what was very wrong: Sherron Watkins at Enron, Coleen Rowley at the FBI, and Cynthia Cooper at WorldCom. None of them meant to be public figures; all did so after their memos to superiors were made public.

Women who see abuses expect that if they bring attention to the situation, a superior with whom they have a good relationship will be supportive and will take it from there. Sherron Watkins at Enron believed this of her boss, Ken Lay, for a long time, as did the other Persons of the Year. Surely, Colleen Rowley thought, the FBI would want to know of lapses made before 9/11. Women in corporations and institutions who advance up the ladder may feel that they are "one of the boys," but upon discovering abuses, the women are separated from the men.

Differences between Men and Women's Expectations

The male perspective, from schoolyard experience on, is that males gang up, take advantage, and that a team player or a guy who wants to stay out of trouble keeps his mouth shut. This "go along to get along" and "look the other way" attitude is a prevailing one in many places.

Something that men instinctively seem to know that women do not is that men in high positions either know what is going on and don't want it known, or, they don't know and would rather not know. In a male world, once a man knows a problem exists, he knows it can make him look bad (be "one down"). Women's experience is different. In the female world, it is not only bonding and stress-reducing to discuss a problem or admit a mistake, it generates support, ideas, and possible resources.

Women are more able to admit mistakes, ask for directions, work collaboratively, seek consensus, be motivated by empathy to alleviate suffering, and take care of the vulnerable. Until women are equal partners in setting values, it is not safe for boys and men to be feeling and nurturing people without suffering from patriarchal judgments that they are not man enough. Patriarchy is endangering all species and the planet itself. Mother Earth and mothers need women to set the priorities, take care of the resources, and stop wars and other abuses of power.

Peace on Earth is Women's Work

Several years ago at Christmas, an email made the rounds, I saw a poster with the same message in a mail order catalog, and then a small framed version came gift-wrapped:

In the Christmas story, if there had been Three Wise Women they would have:

Asked directions,
Arrived on time,
Helped deliver the baby,
Cleaned the stable,
Made a casserole,
Brought practical gifts,
And there would be
Peace on Earth.

My first reaction was to laugh—like most women and some men do. It is humor that depends upon the recognition of how men and women are different, with an assumption of female superiority. It also touched something much deeper in me. Women do have qualities that grow out of being nurturers and caretakers. If there is ever to be peace on Earth, I think it will depend on women bringing these personal and familial skills into the world.

Once it becomes clear to women that we have qualities that are needed for the human family, the planet and all life on it to survive and thrive, the next question becomes, "How do we do this?"

The next chapters point the way.

5

ANTIDOTE 1: THE VISIBLE POWER OF WOMEN TOGETHER

If Mother Earth had a complete physical examination, the tests would come back with reasons to be concerned. The findings: There is widespread pollution of the oceans, rivers, lakes, and streams. The air quality varies and is bad over some major cities. There are fewer and fewer trees and rain forests that serve her as lungs, more extinct and endangered species, less fertile land. Cities, urban sprawl, and the number of people have increased geometrically. The ice caps are melting. Global warming.

Earth is nearing the threshold of overpopulation; soon there will be more people than can be sustained by available resources. Nuclear weapons have metastasized like cancer, which increases the potential for massive destruction of life.

Causative agent: Patriarchy. The result of generation after generation of traumatized children who grow up to be men who repeat the dominator cycle of aggression, humiliation, retaliation and women who cannot protect themselves or their children. Manifests as domestic violence, terrorism, oppression, war.

Needed: An antidote. Definition of *antidote:* "1: a remedy to counteract the effects of poison. 2: something that relieves, prevents, or counteracts (from *Webster's New Collegiate Dictionary*).

Problem: Women holders of the antidote do not realize that *they are the answer and have the power* to prevent and counteract the poison of patriarchy, through the visible power of women together and the invisible power of women in circles.

The Visible Power of Women Together

Stories follow of women who found that they could make a difference when they got together. Often a story is needed for imagination and action. Like starter for sourdough bread, a story generates possibilities in others who rise up, and in turn, a story about what they did becomes the starter for others. Women stand a little taller when they learn what other women have done.

Women are visibly working together at national and international conferences; the most important of them, the United Nations International Conference on Women was last held in Beijing in 1995. The official meetings have developed and passed important resolutions affecting women worldwide.

Whether the antidote is to a local or global problem, whatever is being done anywhere furthers the effort to bring an end to the destructive poison of patriarchy.

Nigerian Activism: Up Against the Corporation

In Escravos, Nigeria, in July 2002, six hundred Nigerian women from ages twenty to ninety took part in a protest, led by a core group who were over forty, and staged a sit-in that shut down a huge ChevronTexaco Terminal. About 150 women managed to sneak into the facility to block the airstrip, helipad, and port that are the only exit routes and held seven hundred workers hostage. The facility is in the oil-rich Niger Delta, surrounded by swamps and rivers. The people who live there are among the poorest in Nigeria, and do not even have electricity. The women staged this protest to persuade the oil company to hire their sons and to give back some revenue to develop their impoverished villages. The women also threatened to take their clothes off, a traditional and powerful damning gesture that collectively shames those toward whom it is directed. Helen Odeworitse, one of the leaders said, "Our weapon is our nakedness." Company officials decided that they wanted to resolve the sit-in through dialogue. They had called in about one hundred police and soldiers with assault rifles, but did not want to use them against unarmed women, including grandmothers and mothers with infants. One woman was roughed up. Taking action was not without risk.

The protest ended peacefully when the company agreed to hire at least twenty-five of their sons; install electricity and water systems in their communities; build schools, clinics, and town halls; and help them build fish and chicken farms so they could sell food to the corporation's cafeteria. Anunu Uwawah, another of the leaders reportedly said, "I give one piece of advice to all women in all countries. They shouldn't let any company cheat them."

By engaging in non-violent civil disobedience, these women forced a multinational corporation to help their villages. They also inspired others to do the same: shortly afterwards, women from other villages occupied four more ChevronTexaco oil facilities in southeastern Nigeria.

When the protest began and before it was ended peacefully, the news of it circulated widely among women through e-mails and Internet news. I was not surprised to read that a core group of older women led the protest. I imagine that this had been preceded by a lot of talk about unemployed sons and poverty in the villages. The huge facility, lit up around the clock, was a self-contained city that brought men in from outside to work in it. Details of the protest were clearly thought out, and as women would do, included provisions of food and the means to cook it.

When women together talk through what to do, their feelings and concerns are brought up and details get covered. Ideas are freely brought up in the circular style of women's conversations. Considering what they did, their talk would have to have focused on the men and their psychology. The importance of work status to their men and how they could fix it must have been a starting point. They obviously took how the men would react to this protest into consideration. When they threatened nudity, they knew the effect on the Nigerian men who would be shamed. They may have guessed that the European men who had the authority to do so would be reluctant and shamed to use force against this large a group of unarmed women, some nursing mothers and others obviously elders. They also knew that they could be physically overpowered and hurt, and they had the courage to do it anyway.

Antidote: Collective genius. Without a structure or hierarchy of power, women meet face to face and talk. This form is a circle. Any circle can become a vessel from which an antidote can emerge. The homey metaphor is of a pot on the stove; the more esoteric is the image of an alchemical vessel. For an antidote to emerge as it did in this example, the problem and its cost has to be expressed and felt at a personal story level before it can become a creative process. Every woman who stays with the process becomes committed to being there, to listen, and to speak up. This becomes the core group. Heat in the form of fear and projections test the vessel. If the vessel holds, the process goes on, drawing upon observations, common sense, imagination, and intuition from each woman. The resulting plan can be, as it was in this case, collective genius. The last step is always courage.

Mothers of the Disappeared

In Argentina, grief and the insistence that they learn what happened to their abducted children made courageous activists out of mothers. From 1976 to 1983, under the rule of a military dictatorship, nearly 30,000 people disappeared. They were kidnapped or arrested, tortured, often mutilated and murdered. The only public protest to this came from the Mothers of the Disappeared, members of the *Madres de la Plaza de Mayo,* who began marching in the plaza in April 1977. China Galland traveled to Buenos Aires in the mid-1990s, when she was gathering firsthand stories for her book, *The Bond between Women: A Journey to Fierce Compassion,* to interview these mothers, who were still marching every Thursday. China and

I were in a women's prayer circle at the time, so I was aware of her research and heard her tell about some of it when she returned. The mothers she met had lost at least one child or member of her immediate family; many had lost multiple family members. Laura Bonaparte, who lost seven members of her family, begged the archbishop to use his influence to help, only to have him reply that she should commend herself to the Virgin Mary and resign herself to her loss.

The mothers began to march in 1977, out of desperation, when no one would help them find out or even be concerned about their abducted children. As China Galland described the situation, "Fourteen grief-stricken, furious mothers turned to themselves, forming the Association of the Mothers of the Plaza de Mayo. In defiance of the military and putting their own lives at risk, they began demonstrating in the plaza. Shortly thereafter, nine of those who had joined them, including a French nun, were taken away by plainclothesmen after a meeting and never heard from again. The Mothers returned to the plaza the following Thursday."

In the beginning, the Mothers of the Disappeared were called crazy ones; later they were considered the conscience of Argentina.

Antidote: Mother courage. The women's movement in the United States had a saying, "Speak truth to power." The Mothers of the Disappeared walked this motto. Every Thursday, wearing photographs of their disappeared family members, they protested by showing up. They marched together in a public place and were visible to others. Men in power who have no problem ordering the use of force against other men, or no problem exercising their personal will or desire on an individual woman, find that they do have a problem when confronted publicly by a group of courageous women, especially if they are mothers and grandmothers. The Mother archetype as symbolized by the Mother Bear is a deep source of strength and courage for women in these situations; the bear is also the symbol of Artemis, the archetype of the sister and protector of vulnerable children and women in Greek mythology.

Nagpur Vigilante Justice

In Nagpur, a city in central India, vigilante justice was as brutal as the rapes that were committed. In three separate incidents in 2004, mobs of women retaliated to rapists with violence. On one occasion, they burned down the houses of three rapists who had reportedly raped with impunity. Earlier, a gang leader, who the women said had terrorized the neighborhood, raped young girls and pregnant women, and then sent his henchmen to extort money, was stabbed and stoned by a mob of women. When five of the women were arrested, more than four hundred women blocked the courtroom and demanded that they be set free. In another case, two men accused of extortion and rape were killed by a mob of women after they tried to strip a local woman. These incidents have set off a public debate about justice. More than a hundred prominent Nagpur lawyers issued a statement saying that the women should not be treated as the accused, but as the victims, and that while it was wrong to take the law into their own hands, the courts should look into the circumstances.

It was front-page news in Nagpur and other parts of India. I read the story in an American newspaper. That women were raped, especially poor women, is not newsworthy. That women together would rise up and use violence to stop the rapes was news. This would certainly discourage other would-be rapists in Nagpur and make the streets safer for women.

Their wrathful energy was a collective expression of the archetype of Kali—one of the representations of the "enough is enough" archetype, and a Goddess of Transformative Wrath, which I described in *Goddesses in Older Women.* The Hindu goddess Kali is a fierce protectress, terrifying and bizarre in appearance, who like Sekhmet, a goddess of ancient Egypt with the head of a lioness and the body of a woman, came forth when evil became too much for the gods to overcome and only a goddess could succeed.

It is significant that in mythologies such as these, when evil became too much for the gods or men to contain and overcome, the mythic response is to call forth the feminine. The evils of war, terrorism, fratricide, and the need to dominate are too much for men as a gender to overcome, since war is sanctioned by patriarchal religions and hierarchical conditioning shapes men's allegiances.

Antidote: Women's wrath. The Nagpur protest against men raping with impunity became mob violence. Outrage got out of control, which is also in the mythologies of Kali and Sekhmet. If rage takes possession of either a woman or a mob of women, the effect is the same: head and heart are out of action for the duration. The energy itself may, however, be needed to bring about change. The intensity of this archetype needs to be focused, harnessed, and balanced by a circle of women to be used most effectively.

Date Rape/USA

In the United States, date rape can be done with impunity because it is her word against his. Besides, she did go out with him and was alone with him when he did it. In one such situation, women friends went to the date rapist's office and created a scene to inform his coworkers of his private behavior. With signs and chants, they made his behavior public. With truth the best defense against libel, the shoe of proof was now on the other foot. This was just desserts and a light form of vigilante justice. This public demonstration also placed shame where it belonged rather than on the woman who usually suffers twice—once from the rape, and then from the shame that it was done to her.

Antidote: Sisterhood support. A circle of friends heals the shame of the victim, and by taking effective action, particularly suited to the situation, justice is done.

UCSF Women's Seminar

Women who get together, know, and trust each other become a sisterhood of support and action. The date rape response is one example. Women's seminars, support groups, or regular get-togethers of any kind can also rise to the occasion when needed.

When I led a seminar for women psychiatry residents at Langley Porter Psychiatric Institute, University of California San Francisco (UCSF), a resident who thought that she must have somehow invited the inappropriate sexual advances of a supervisor spoke about the incident and found that two of the other women in the seminar had similar encounters with the man. Since women use conversation to bond, and sharing similar experiences and feelings are what friends do to stay equals, once the seminar had become a closed vessel where confidences would be held, it became safe for her to speak and for others to share.

That women have had similar experiences is not surprising; that two others in the same seminar had the same experience with the same supervisor made it unusual. From their respective stories and the discussion that followed, it became apparent that his behavior would not have been inappropriate in a social setting, and once his advances were discouraged, he hadn't persisted. He seemed unaware that his supervisory role made his behavior not just inappropriate, but reportable. The three decided not to report him but confront him together. They thought that he was a good man who had something important to learn, for his own sake and for other women he would supervise. They believed that he would stop this behavior as a result of their intervention. They also decided to warn incoming women residents and to wait and see.

Once women in any institution trust each other and share information about whom to be wary of and why, or decide to confront an individual or report bad behavior, the workplace is safer. Once a department or an organization is aware that women get together regularly and keep what is said in confidence, the women do not have to take any actions or confront anyone to have "gained some muscle."

In this same seminar, another very serious ethical violation of the doctor-patient relationship came up later. A woman in the seminar told us that her former psychiatrist, who was a faculty member, had had sex with her in her sessions; she still was coping with the damage that it had done to her. She had learned that he was now about to be honored, and had decided that she would go to the chair of the Department of Psychiatry to tell her story.

As a young resident, going into his office entirely on her own, especially if her former psychiatrist denied it, put her at risk of not being believed, which would have compounded the situation for her. We discussed ways that we could become involved beyond providing moral support. Did she want others to go with her? Her response to this was, No, it was important to her that she do this herself. The solution was to write a letter to the department chair, telling that she had discussed the violation in the seminar and that we all believed her. The letter was signed by me and by the coleader, both of us respected faculty members. After listening to her story and receiving our letter, the chairman called her former psychiatrist into his office, told him that he believed the resident, and that the psychiatrist could either resign or come before an ethics committee that would look into the charges. The psychiatrist resigned.

Antidote: Sisterhood support. If this had been a women's circle with a spiritual center, there would have been an additional dimension. A silent meditation, perhaps, to tap into the wisest course to take. A silent prayer or a ritual, perhaps, to bless her and give her courage to tell her story and be believed. If there had been e-mails then, we would have all been informed of the time of the appointment, and been mindful and sent her invisible support.

Women's Sanctuary Forest

There is a fourteen-acre grove of old-growth, magnificent redwoods on the edge of the Mattole River in Humboldt County, California, about five hours north of San Francisco. Some are more than a thousand years old. They will not be logged, because they are now in a permanent land trust. Activism took the form of making a commitment to buy the land. Catherine Allport saw the possibility, talked with her friend Tracy Gary, and in 1994 they started the Sacred Grove Women's Forest Sanctuary, a nonprofit organization whose mission is to save a stand of old-growth redwoods by buying the acreage and keeping it from being logged in perpetuity. Other women joined them in a core circle to raise money to pay the mortgage, which is due four times a year. They gather together once a month on a Sunday afternoon and meet first as a circle—

checking in, staying current with each other's lives, observing special times, being silent together, making beauty and ritual—doing what women's circles do that is soul nourishing. Then they turn their attention to what needs to be done for the trees, which may be about raising money for the next payment, doing a mailing, holding a fund-raising or a thank-you event for donors, or sponsoring an event or workshop.

Antidote: Financial activism. When what needs saving can be purchased, a commitment to do so by women who take on the responsibility to raise the money together gives back to them psychologically. They will do what needs to be done, and in doing so, friendships deepen and new tasks are taken on.

Ms. Magazine

Founded in 1972, *Ms.* magazine was the publication most responsible for raising the consciousness of women about sexism and inequality, without which it is unlikely that the 1970s would be called the "the decade of the women's movement." The magazine had financial difficulties because it lacked an income base from advertisements. While never relinquishing editorial independence, the magazine was sold and resold three times between 1987 and 1998 and was at risk of stopping publication.

Antidote: Financial activism. *Ms.* magazine was purchased in 1999 by fourteen feminists with the financial means who had been brought together by Gloria Steinem and Marcia Gillespie to form Liberty Media for Women for this purpose. They then transferred ownership to the Feminist Majority Foundation.

According to the Federal Reserve, women control 51.3 percent of the private wealth in the U.S., with more on the horizon. The largest wealth transfer in history is about to take place as the Baby Boom generation inherits from their parents. Forty-one trillion dollars are expected to pass from generation to generation in the next fifty years. Since women generally outlive their husbands, the family assets will become concentrated in the hands of Boomer women, who are expected to survive their husbands by fifteen to eighteen years.

Philanthropic choices will make a difference. Since its founding in 1987 by feminist leader Eleanor Smeal and philanthropist Peg Yorkin, for example, the Feminist Majority Foundation has not only bought *Ms.* magazine, it has also supported many national and international efforts to improve the lives of women and girls. When I was on the board of the Ms. Foundation for Women in the 1980s, this still-going-strong foundation was the only national foundation at the time to support women and girls.

Material wealth is like good health, good genes, and natural talent—it is part of our personal, spiritual, and psychological journey to use or to waste. I think that to whom much has been given, much is expected. I also know that the first step is usually appreciation for what one has. Using what we have been endowed with, metaphorically and materially, makes life more meaningful. To become wise and able to take compassionate action will be the challenge that comes to many Baby Boomer women in this third phase of life when "What do

I do with my money?" is one question and "Who do I become through my choices?" the other.

Rwanda Mothers

The genocide in Rwanda was horrible and appalling. Initially, 800,000 people were killed in the first one hundred days by militant Hutu who systematically murdered the Tutsi and moderate Hutus. More died later. Tutsi women were targets of rape and mutilation. Hutu women in mixed marriages were raped as punishment and forced to kill their Tutsi children. The report issued by the Organization of African Unity said that virtually every surviving female over the age of twelve had been raped. Raped women gave birth to from 2,000 to 5,000 "children of hate." In a patrilinear society, the children would be considered Hutu, which was one of the motives. When the genocide ended, there were three million refugees, 120,000 people in prison, and 500,000 orphans. The social structure had been destroyed, and in every community, the women of Rwanda were left to clean up and reconstruct what was left.

Antidote: Women's reconciliation workshops. Aloisea Inyumba was appointed to manage the aftermath as minister of family and women's affairs. Her task was to help the living find a way to begin again. Inyumba started the Rwanda Women's Initiative, a national women's grassroots network modeled on the Bosnia Women's Initiative, which brought Hutu, Tutsi, and Twa or pygmy women together to talk about their common needs. She described the weeklong workshops for reconciliation that were held; at the end of the workshops the women opened up to the possibility of working together. One of their first concerns was for the orphans. "Women had lost children, children had lost mothers. We did a national campaign, we said, 'Every home a child, every child a home.' Women went to the orphanages and took children home, Hutu and Tutsi women have all taken children, regardless of ethnic background. It was the first step in reconciliation."

Antidote: Women as peacemakers. Women as a gender can talk and listen to each other with empathy, bonding, and an increase in oxytocin. The form most suited is a conversational circle. This would be especially beneficial in parts of the world where there have been generations of fratricidal conflict between men. Mothers and grandmothers who want peace could get together to know one another and become a force for peace. In Northern Ireland, during thirty years of "troubles" between Catholics and Protestants, small groups of women on each side of the conflict did meet together, as they are doing in Jerusalem today.

Bonobo Females

Women could learn from bonobos. As Natalie Angier describes,

> There is one creature that stands out from the chest-thumping masses as an example of amicability, sensitivity, and, well, humanness: a little known ape called the bonobo, or, less accurately, the pygmy chimpanzee.

> Before bonobos can be fully appreciated, however, two human prejudices must be overcome. The female bonobo is the dominant sex, though the dominance is so mild and unobnoxious that some researchers view bonobo society as a matter of "co-dominance," or equality between the sexes. . . . The second hurdle is human squeamishness about what in the '80s were called PDAs, or public displays of affection.

Proponents of male dominance as the natural order point to chimpanzee behavior as evidence. These genetically close relatives of humans are aggressive, prone to violence, invested in hierarchy, hunting, warfare, and male dominance. Dr. Frans De Waal, an expert on primates, makes the point that bonobos refute the idea that we can't help but behave like our chimpanzee primate ancestors.

The bonobos are equally close to us genetically (both species share at least 98 percent of human's DNA), and yet behave very differently. Bonobos would rather make love than war. They use sex to appease, bond, make up after a fight, and to release tension. Sex is casual and quick, an everyday social interaction. Every bonobo is sexually free to do whatever with whomever, whenever inclined.

De Waal describes female bonobos as forming "constructed sisterhoods," which gives them an edge over males because they stick up for one another. If a male acts aggressively toward a female, other females will come to her aid. He speculates that it could be that female alliances arose to prevent infanticide by males, which is common among chimpanzees and other species, but has never been observed among the bonobos. Females form strong alliances with other unrelated females. Bonobo adolescent females disperse, which prevents incest. In effect, they leave home, move to a new community, make new friends, and become part of a sisterhood.

Together bonobo females ensure their own safety and the safety of their young. Though a female is smaller and has less physical strength than an adult male, she does not need to fear, because females will come to her aid if any male tries to overpower her or take her child. Bonobo society is one in which males are not invested in hierarchy, where there is little if any everyday violence, no war, and no fighting over sex or possession of females or rape.

Antidote: Women's support circles. Taking a lesson in sisterhood from the bonobos, women who meet together in a support circle can find ways to come to each other's aid. Cell phones make communication easy. Defusing a situation before it builds up might be that friends show up on the scene, or there might be a standing offer of a place to stay if a need arises for a safe harbor. Women in a circle share information about steps to take and resources that are available, and tell what has worked for them.

Antidote: Group protection. In a village in Uganda, a woman was regularly and severely beaten by her husband. Like many abused wives, she said that she deserved the beating when other women voiced their concern. Only when they pointed out that if he killed her, her children would be without a mother did she agree to let them help her. A plastic child's whistle was the solution. The next time he began to beat her, she blew the whistle. The women in nearby huts heard the whistle and came running into the hut. Each said to the man, "Beat me, beat me." Confronted in this way, the wife beater didn't beat any of them. Soon all the women in the village carried these little penny whistles, and none of them got beaten. The penny whistle idea was then passed from one village to another, which stopped wife beating and empowered women. The penny whistle movement eventually led to legislation that made spousal abuse illegal in Uganda.

Activism: The 9/11 Widows

Mindy Kleinberg, Kristen Breitweiser, Patty Casazza, and Lorie Van Auken—all four were mothers widowed by 9/11. They came together and sought answers and explanations, and pushed for the establishment of an independent commission, which was opposed by President Bush, to investigate 9/11. "The Jersey girls" is what they were called in Washington. They persevered, staged a Washington rally, buttonholed legislators—boarding elevators that said "Senators Only," prodding Congress and a recalcitrant White House—to create the panel. Once it was formed, the widows lobbied for a bigger budget, and got Condoleezza Rice to testify publicly and the president and vice president to at least testify behind closed doors.

"They call me all the time," said Thomas H. Kean, the commission's chairman and a former Republican governor of New Jersey. "They monitor us, they follow our progress, they've supplied us with some of the best questions we've asked. I doubt very much if we would be in existence without them." The story of how they moved a seemingly immovable bureaucracy was "a tale of a political education and a sisterhood born of grief."

Information that would otherwise not have been made public indicated that there were no Iraq-9/11 links, and that the attack on the World Trade Center could have been averted if the FBI had paid attention to information from field reports and from warnings that hijacked airplanes could be used as weapons. The panel concluded that more attacks were likely, and made forty-one specific recommendations, which the widows of 9/11 were determined would not be ignored.

Antidote: A core circle. This foursome was an exceedingly effective core group for activism. A small circle of women can stay in touch by phone and e-mails, meet together, and share discouragement and elation. Women's courage and ability to speak out and not get flattened by resistance to hearing them comes from being in it together. It is also sustained by knowing that whatever they are doing may (in their case) save lives or help others, which taps into the women's maternal concern.

Years ago, when Gloria Steinem was on the front lines (when has she not been?) and had negative projections and hostile remarks directed at her, I assumed she had a spiritual source that sustained her. I was seeing other feminists burn out, while she did not. Back then, *spiritual* and *feminist* were rarely used together, at least not on the East Coast, where feminists were intellectual and political. After fiddling with how to frame the question, I got the answer. Gloria was the spokesperson; she went out into the patriarchal world that was threatened or ignorant. But she always came back to the *Ms.* magazine office, where everyone was working for the same vision, each doing her part. She was the point person and they were all on the same team. They were friends doing meaningful activism.

Whether it is the Mothers of the Disappeared in Buenos Aires, the 9/11 widows, the *Ms.* magazine office, or women saving a forest, women who work together to make a difference take courage from each other, are motivated not only by the cause, but on a daily level keep the course because they do not want to let each other down.

The 30 Percent Goal

WEDO—the Women's Environmental and Development Organization calls attention to the United Nations' critical figure of 30 percent that is needed to maintain the impetus toward equal female/male representation in government. In the 2005 report, fifteen countries reached or surpassed the 30 percent goal in their national legislatures. Eleven were in Europe: Sweden (45.3 percent), followed in descending order by Denmark, Finland, Netherlands, Norway, Spain, Belgium, Austria, Germany, and Iceland; three were in the western hemisphere: Cuba, Costa Rica, and Argentina; and two were in

Africa: Rwanda after the genocide has 48.8 percent women in their parliament and South Africa has 32.8 percent.

By comparison, women comprise only fourteen percent of the 109th United States Congress (2004–2006). There are sixty-nine women in the House of Representatives and fourteen women in the Senate, and not all of them can be counted on to have strong sister or maternal concerns for women, children, or peace.

Antidote: Political activism. In 1985, twenty-five women gathered in Ellen Malcolm's basement to send letters to their friends about a network they were forming to raise money for pro-choice Democratic women candidates. At that time, no Democratic woman had been elected to the United States Senate in her own right, no woman had been elected governor of a large state. These "founding mothers" created a new concept in fundraising: a donor network that would provide its members with information about candidates and encourage them to write checks directly to the candidates they chose. EMILY's list has become America's biggest political action committee. EMILY is an acronym for Early Money Is Like Yeast. (It raises the dough.)

Women with maternal values of caretaking and peacekeeping are needed to run for office and be elected in sufficient numbers to influence legislation, appointments, and money spending. Otherwise concern for children, freedom of choice, and freedom from violence take a backseat to alpha male agendas.

For women to enter politics is daunting. It may, however, be a personal "assignment," something that you feel drawn to and think you can do. Others may try to talk you out of it with the same arguments that girls and women used to hear when they said they wanted to be doctors. It is a long-term commitment that will take personal courage and sacrifice, ambition, and a desire to make a difference. Marie Wilson saw the need to set the goal high when she called the organization she founded "The White House Project" and focused on getting women involved in politics. Its Vote, Run, Lead initiative teaches young women how to organize, how to get out to vote, and how to run for office.

Medical and law schools regularly admit classes now that are 50 percent women. The time needs to come and soon, when the Congress of the United States looks like the population it represents.

Beijing 1995

The Fourth World Conference on Women, Beijing, 1995, was the largest-ever gathering of the world's women. There were more than 50,000 participants. Five thousand delegates from 189 member states attended the official United Nations conference, with an additional 4,000 representatives from nongovernmental organizations (NGOs). The NGO Forum, which preceded the UN Conference, had more than 30,000 delegates. These particular NGOs are those that serve women and girls. The conference turned out to be a women's conference about the state of the world as well as about the status of women in the world.

Most significant, *women's rights were recognized as human rights* in the Beijing Platform for Action. Until this conference, women were considered passive victims in an unequal world. The conference addressed *gender relations* rather than women's issues, recognizing that women's roles and status are in relation to men. Governments agreed to promote gender equality and women's empowerment in twelve areas of concern, and to report progress in 2000 and 2005. These areas were: the burden of poverty on women, unequal access to education and training, inadequacies and inequalities of healthcare, violence against women, protection of women in conflict situations, inequalities in economic empowerment, inequalities between men and women in power sharing and decision making, insufficient mechanisms for advancement of women, lack of respect and inadequate protection of human rights of women, media stereotyping and inequality of access to communication, gender inequalities in the management of natural resources and safeguarding the environment, and persistent discrimination against and violation of the rights of the girl child.

The word *gender* was challenged in Beijing by countries that do not see women and men as equal human beings but as necessarily different sexes with fixed social roles. This challenge was overcome.

More of This Antidote Needed: A Fifth World Conference on Women

In 2002, I attended meetings at United Nations Commission on the Status of Women as a member of a contingent from the

Millionth Circle, and learned that there were *no* plans to hold another international women's conference. It was a shock because of the assumption that there would be one in 2005, ten years after Beijing. Previous conferences had been held regularly since 1975. Each had been larger and more significant than the one before.

With a three-year lead time needed to plan and make arrangements, there definitely would not be a fifth conference in 2005. There would not be one at all, ever again, unless new momentum could revitalize it. The United Nations was deeply involved in events following the terrorist destruction of the World Trade Center on September 11, 2001, and on the question of whether Saddam Hussein had weapons of mass destruction, as well as the conflict between Israel and the Arab world and other ongoing fratricidal wars and aftermaths. War and aggressive posturing took central stage, as usual. There also were fears that a fifth world conference, led by the American delegation's agenda, which would be that of the Christian fundamentalists, might dismantle the Platform for Action.

At every previous conference, women came from all over the world, talked and bonded, learned from each other, and formed a network of friendship and political alliances on behalf of women. Since Beijing, we have the Internet as a means of communication and connection. It gives us power at our fingertips to move ideas and funds, to mobilize and organize and make a difference. E-mails would facilitate meetings at a Fifth World Conference, and make follow-up and continuation easy. Listservs would make distribution of information immediate;

Web sites and links among them would be invaluable. The value is getting things done through personal alliances and connections. Finding out who to reach and how is often a matter of one to three degrees of separation. More would be done for women by women.

Antidote: One of the goals envisioned by Millionth Circle at the United Nations is to seed the idea of circles with a spiritual center as a source of support and an antidote to burnout. Loss of meaning, indifference, and even cynicism in formerly idealistic and caring women is an occupational hazard. It comes from overwork and frustration, trying to help people on one hand, and deal with indifference or incompetence or lack of funds on the other. Women's NGOs are on the front line, trying to alleviate suffering caused by wars, famine, and exploitation of women and children. When women can tell others in a circle what is going on in their lives—personal as well as at work—they may shed tears, express anger, frustration, feelings of vulnerability, set off peals of belly laughter, or be prayed for, with the contents held in confidence. A circle of women is a nurturing and sustaining resource that can become a spiritual and psychological wellspring tapped into whenever the circle meets, whether in an office building in Manhattan or anyplace in the world.

Lessons from Geese

The underlying principles of being in a functional circle of people working on a common goal has everything in common (except the shape) with what geese do when they fly in formation. "Lessons from Geese" was transcribed from a speech given by Angeles Arrien at the 1991 Organizational Developmental Network, and was based upon the original research of naturalist Milton Olson.

Fact 1: As each goose flaps its wings it creates an "uplift" for the birds that follow. By flying in a V formation, the whole flock adds 71 percent greater flying range than if each bird flew alone.
Lesson: People who share a common direction and sense of community can get where they are going quicker and easier because they are traveling on the thrust of one another.
Fact 2: When a goose falls out of formation, it suddenly feels the drag and resistance of flying alone. It quickly moves back into formation to take advantage of the lifting power of the bird immediately in front of it.
Lesson: If we have as much sense as a goose, we stay in formation with those headed where we want to go. We are willing to accept their help and give our help to others.
Fact 3: When the lead goose tires, it rotates back into the formation and another goose flies to the point position.
Lesson: It pays to take turns doing the hard tasks and sharing leadership. As with geese, people are interdependent on each other's skills, capabilities, and unique arrangements of gifts, talents, or resources.
Fact 4: The geese flying in formation honk to encourage those up front to keep up their speed.

Lesson: We need to make sure our honking is encouraging. In groups where there is encouragement, the production is much greater. The power of encouragement (to stand by one's heart or core values and encourage the heart and core of others) is the quality of honking we seek.

Fact 5: When a goose gets sick, wounded, or shot down, two geese drop out of formation and follow it down to help and protect it. They stay with it until it dies or is able to fly again. Then, they launch out with another formation or catch up with the flock.

Lesson: If we have as much sense as geese, we will stand by each other in difficult times as well as when we are strong.

I had fun and a great reception when I spoke at the Quw'utsun Cultural Center at Malapina University College-Cowichan Campus, in Duncan, British Columbia, about *The Millionth Circle* and read "Lessons from the Wild Geese" (I took the literary liberty of differentiating those in the sky from the white geese I see around ponds). Wild geese are Canadians, after all. I added a one-line song. Appropriately, it's sung in a round: "There's a river of birds in migration, a nation of women with wings" (composer unknown). And for good measure, I added a circle cheer of support—*Honk! Honk! Honk!*—and suggested that instead of a secret handshake, they could flap arms as if in flight. Activists do need to laugh together and take themselves lightly while changing the world.

6

ANTIDOTE 2: THE INVISIBLE POWER OF WOMEN'S CIRCLES

The invisible power of women's circles acts on two levels. A circle of women may appear to be just women talking. But if it is a circle, especially one with a spiritual center, it will have an invisible effect on the women in it. A second invisible power is the possibility that each circle is contributing to a critical mass that will bring an end to patriarchy. This is the effect on culture of *the* millionth circle.

The power to resist "Who do you think you are?!" criticism or disbelief from outside sources comes from being in a circle with like-minded others, which allows women to keep on course in the face of ridicule or opposition. The invisible power of women's circles on the women in them grows out of the power that we have on one another, which can be healing, affirming, and supportive. Whenever there is encouragement and practical support to make a significant change, change is more likely to happen. That others believe in us or have the same perspective we have has a powerful and invisible effect.

A circle of women who trust each other can also become a vessel of healing, especially when women in them are able to talk of experiences in which they were terrorized by what was done to them or what they witnessed. A safe place to tell the truth is a healing space. An abused person is twice emotionally wounded: by what was done and by shame. Deep in the psyche of every abused or raped woman and child, there is shame and a sense of unworthiness and rejectability. This is made infinitely worse when religion calls these victims sinful. Every time that a woman musters the courage to speak and it proves to be safe, trust grows and her psyche gradually heals. When others listen with compassion, invisible wounds and scars gradually heal.

A circle is also an experience in egalitarian conversation that can be carried over into other relationships. In a circle, the habit of being able to express ideas, needs, feelings, and hopes develops, as does listening. In a sense, being in a circle is a practice in paying attention and receiving attention. In a functional circle, no one person dominates. Our significant relationships are either hierarchical (there is an unspoken agreement that one person's opinions, needs, feelings, and perceptions are more important than the other's, which is a patriarchal model) or a circle (both speak and listen to each other, as equals who are important to one another). Recalling Eleanor Roosevelt's words, "No one can treat you as an inferior without your permission," the task often becomes transforming your piece of patriarchy into a circle. The invisible effect of being in a circle makes this possible.

The Millionth Circle Idea

The Millionth Circle as an idea started to grow in my mind when I was in an imaginary rain forest in Grace Cathedral in San Francisco. I was there as a guest of members of an organization called the Foundation for Global Community, and was remembering when they were known as Beyond War. Projected against the walls and columns of the cathedral were images of a tropical rain forest, with recorded sounds of the rain forest in the background. Indigenous spiritual leaders spoke in turn in their native tongues and were translated. They spoke as if in a circle, from each of the four directions. The combination of these elements made it a timeless, meditative experience, where thoughts and memory gathered.

I remembered when Beyond War had been a most effective organization that worked to stop the proliferation of nuclear weapons. Unlike other organizations with the same goal, they did not emphasize the looming image of a mushroom cloud and destruction. Beyond War focused upon images of the beauty of the Earth. The music of "America the Beautiful" was the sound track of a film full of images to match the words—of spacious skies, amber waves of grain, and so on. There were faces of the people and of the children of the world, the human family in its diversity. It was a *call to action to save what we loved.* Beyond War provided an education about what could be bought with money we Americans were spending on armaments, one item at a time. The cost of one tank, or one missile, or one battleship and what it could otherwise buy to keep children healthy, build schools, or open clinics. Amazing, the civilian purchasing power represented by each war object. I remember that one battleship was then worth all of the campuses of the University of California.

The day before, I had been writing about the archetype of the circle and circles of wise women for *Goddesses in Older Women,* so circles were in the background of my mind. Beyond War was also a reminder of "The Hundredth Monkey," the allegory that kept antinuclear activists believing that they could have an influence on the nuclear arms race between the two superpowers, when conventional wisdom thought this absurd and even if no immediate effect could be seen. Then I overheard someone mention "a million signatures," and the idea of "the millionth circle" came fully into my mind. A process akin to chemistry lab: you have a beaker filled with a solution in which you have dissolved a number of chemicals, and there is nothing to be seen until one last something is added, and a whole crystallized form precipitates out of the solution into visibility. The millionth circle, like the hundredth monkey, is the one that tips the scales. Or it is like that last something added to a beaker of solution. It is the circle that, when added to the rest, causes a new form to emerge and bring an end to patriarchy.

The Millionth Circle: How to Change Ourselves and the World—The Essential Guide to Women's Circles was the result. It is a thin book with a minimum of words. I told the story of the hundredth monkey and a short explanation of critical mass, and followed this with chapters on how to create circles with a sacred center, a "Zen and the Art of Circle Maintenance." The little book went out in the world and has functioned as a seed packet to grow circles and inspire others to form programs and create new organizations.

Circles with a Center

A circle with a spiritual center invites the invisible world of soul and spirit to be in the center of the circle and in each person in the circle. One image is of people sitting around an invisible campfire that is a source of warmth and light, aware that they also have a similar source inside them. Or it is like a wheel with spokes connecting points on the rim to the center. There is no hierarchy. Through meditative silence and silent prayer, wisdom and compassion can enter to center us. Anything that signals that the circle has begun can shift the energy from social to sacred.

When the circle is a place where others listen with empathy, where judgment or comparisons are not made, and whatever is said in confidence is held in confidence, it is a circle with a spiritual center even if not recognized as such. If there is love and trust, the circle is a sanctuary. If it is a circle of women, there will be enough Mother Nurture to go around. Women's circles become a womb space, where dreams and plans are incubated, and a place to voice them and be supported to take our first steps. The explicitly spiritual may not come into a circle until someone's health or child is in need of prayer, and yet it is a sacred circle. When fierce compassion and concern for justice is the focal point of a circle, that circle will energize the women in it and is a circle with a center. I think that meetings of the *Ms.* magazine staff, the 9/11 widows, or the core group of Nigerian women who organized the protest against ChevronTexaco had the egalitarian structure of a circle and the energy of a circle with a center.

Critical Mass: A Tipping Point

How does a conviction initially held by relatively few people spread through a population to become generally held? How did ideas discussed in consciousness-raising groups gain momentum and become a movement? Malcolm Gladwell in *The Tipping Point* explains that this happens when there is a critical mass. His is an epidemiological model: An idea can become contagious and spread like a virus, through geometric progression by doubling and doubling again, and again and again, until it reaches a critical mass, which is the tipping point. If we are talking about viruses, the result is an epidemic. If we are talking about changing the patriarchal pattern to an egalitarian principle, and a culture based upon balance between masculine and feminine qualities, we are talking transformation. When a critical number of people accept a principle, it becomes the new standard, an "as if it always was so." Like voting rights for American women, for instance, which we now take for granted.

Geometrical Progression

The example that Gladwell uses to make his point about geometrical progression is truly mind-boggling. He asks that we consider the following puzzle: "I give you a large piece of paper, and I ask you to fold it over once, and then take that folded paper and fold it over again, and then again, and again, until you have refolded the original paper 50 times. How tall do you think the final stack is going to be?" No one guesses this right. Answer: If it were possible to do, "the stack would be as high as the distance to the sun. And if you folded it over one more time, the stack would be as high as the distance to the sun and back."

When I thought about how women's circles with a spiritual center would proliferate, geometrical progression was not on my mind—strawberry plants were. It seems far-fetched to equate the two, and yet, the principles are much the same. Strawberry plants send out runners in all directions, and wherever these viney outgrowths touch down, a new plant can grow. Every women's circle is like a strawberry plant, and any woman in one can get enthusiastic, tell others and plant the idea, and another circle forms, and so on, until the field is full of strawberry plants or the world is full of women's circles.

The tipping point occurs when a critical number of people embrace a new idea, value, or perception. Before Rosa Parks, white people expected black people to obey the signs that said "Whites Only" and know their place as inferiors. Her decision to refuse to give up her seat was a personal one, and yet her timing evoked a collective response. A significant number of black people were fed up with how they were treated, especially after many of them had lived in other parts of the United States or been in the armed services after it had been desegregated. Her arrest was the tipping point event. Martin Luther King Jr.'s speech gave words to her action, and their feelings: "There comes a time that people get tired. We are here this evening to say to those who have mistreated us for so long that we are tired, tired of being segregated and humiliated, tired of being kicked about by the brutal feet of oppression." For the black community of Montgomery, boycotting the city bus system in protest was an idea whose time had come. Desegregation became a contagious idea that spread beyond Montgomery to become the civil rights movement.

The bus boycott led to sit-ins and arrests at restaurants that refused to serve blacks; there were marches in protest of segregation and strife over voter registration. Numbers of black people were jailed and beaten, three white civil rights workers were murdered, black churches were torched. Civil rights legislation was passed and desegregation of schools occurred with federal marshals walking children between lines of jeering, hostile adults. All of the "Whites Only" signs came down.

Social Epidemics

Malcolm Gladwell makes the point that social epidemics behave like their disease equivalent and take a small percentage of the population to bring about. Epidemics depend upon the people who transmit infectious agents, the strength of the infectious agent itself, and the susceptibility or resistance to it. Social epidemics work in the same way; it matters who spreads the idea, that it takes hold, and the receptivity of people to it.

Using Gladwell's criteria, for there to be a millionth circle tipping point, the idea has to be spread by three types of people:

by some who are enthusiastic and energetic, are widely known, and are held in high regard by their peers; by others whose knowledge is valued and who pass on information about the millionth circle *with the sole purpose of wanting to help others;* and by still others who sell the idea and overcome resistance. All types need to believe that change and transformation is possible through circles, and want to make a difference through what they are doing to further the goal of reaching a critical mass. As Gladwell says in *The Tipping Point*,

> What must underlie successful [social] epidemics, in the end, is a bedrock belief that change is possible, that people can radically transform their behavior or beliefs in the face of the right kind of impetus. . . . In the end, Tipping Points are a reaffirmation of the potential for change and the power of intelligent action. Look at the world around you. It may seem like an immovable, implacable place. It is not. With the slightest push—in just the right place—it can be tipped.

Millionth Circle, Critical Mass, Morphic Field Theory

Gladwell's tipping point and critical mass are ideas easily grasped by logic. In contrast, the relationship between critical mass and morphic fields, which theoretical biologist Rupert Sheldrake proposed, requires intuitive deduction from evidence which—like Darwin's theory of evolution—is convincing to some, and not adequately proven for others. It is the theory on which I based *The Millionth Circle:* Every species has its own morphic field, through which all members of the species are influenced and in turn affect. The collective unconscious that C.G. Jung named and described is the same as Sheldrake's morphic field.

In Sheldrake's theory, as the millionth circle movement grows through the formation of new circles, it will draw upon the energy or patterns of similar present or past circles. These could theoretically be suffragette circles, consciousness-raising groups, Alcoholics Anonymous or other recovery meetings based on AA that call upon higher power, indigenous councils, Quaker meetings, and even circles that human beings held before there was patriarchy. Sheldrake's theory of morphic resonance is that each new circle will draw from all the circles in existence before it, and in turn will contribute to this same field. The more circles there are, the easier it is for still more to form, which increases the momentum as a movement grows, until a critical mass tips the scales and changes the behavior of the species. Sheldrake's theory explains how new or learned behavior can become a natural or instinctual way of doing something once a critical number of the particular species does this new thing.

For example, in *The Presence of the Past,* Sheldrake describes new behavior by bluetits, small birds that usually fly only a few miles from their breeding grounds. These birds learned that the milk bottles left on back doorsteps in the morning had tops that they could tear with their beaks and suck up the milk on top. First noticed in Southampton in 1921, this behavior spread through the British Isles and Europe. Then during World War II, milk deliveries stopped, which meant that several generations of bluetit birds would come and go without ever seeing a milk bottle. As soon as milk deliveries were resumed, however, bluetit birds were at it again, tearing off milk bottle tops and sucking up the milk.

Form Circles and Change the World

The circle is a form that women take to immediately and often spontaneously when they arrange the furniture for a meeting. The hierarchical arrangement is in rows that face the authority. People in rows can't see each other. Women usually defer to the traditional form when it is in place, though more and more, furniture is rearranged and meetings are held in circles, even in hotel conference rooms. In circle, people face each other and are in equal relationship to each other.

Sheldrake adds the morphic field as a dimension through which circles in general or women's circles with a spiritual center in particular are building toward a critical mass. The human morphic field is the collective unconscious. Every circle formed anywhere in the world draws from and contributes to this archetype. A circle formed in San Francisco will make it easier for a circle in Kabul to form. A microcredit unit of five women in Bangladesh contributes to the formation of a new circle in Kansas City. Each circle is activating the archetype, and each new one is making a contribution of itself to the millionth circle. While we cannot know how much we contribute to furthering momentum toward a critical mass, knowing that we are doing so adds a sense of being connected to others all over the world and being part of something larger.

In "Mis estimados: Do Not Lose Heart," an essay that can be found on the Creative Resistance website, Clarissa Pinkola Estés wrote:

> Ours is not the task of fixing the entire world all at once, but of stretching out to mend the part of the world that is within our reach. Any small, calm thing that one soul can do to help another soul, to assist some portion of this poor suffering world, will help immensely. It is not given to us to know which acts or by whom, will cause the critical mass to tip toward an enduring good. What is needed for dramatic change is an accumulation of acts, adding, adding to, adding more, continuing. We know that it does not take "everyone on Earth" to bring justice and peace, but only a small, determined group who will not give up during the first, second, or hundredth gale.

"A Tale for All Seasons" about "nothing more than nothing," by Kurt Kauter came in an e-mail, just in time for me to add it here. Whether a snowflake or the hundredth monkey or the millionth circle, the message is the same: keep on doing what you believe in. Don't stop because you cannot see the difference you are making. Keep the faith that when you are kind, or see that justice is done, or make anyone in this world happier or any place on Earth more beautiful, that you are contributing to peace.

> "Tell me the weight of a snowflake," a sparrow asked of a wild dove.
>
> "Nothing more than nothing," was the answer.
>
> "In that case I must tell you a story," the sparrow said. "I sat on the branch of a fir, close to its trunk, when it began to snow. Since I did not have anything better to do, I counted the snowflakes settling on the twigs and needles of my branch. Their number was 3,741,952. When the 3,741,953rd snowflake dropped onto the branch, nothing more than nothing, as you say, the branch broke off."
>
> Having said that, the sparrow flew away.
>
> The dove, since Noah's time an authority on the matter, thought about the story for a while, and finally said to herself, "Perhaps there is only one person's voice lacking for peace to come to the world."

7

ANYTHING WE LOVE CAN BE SAVED

I think that it is time to acknowledge that women as a gender—as a whole, not every woman, but women generally—have a wisdom that the world needs now. Lots of women are already doing their part. It was my assignment to write this book. By "my assignment," I mean that the idea came to me and I grabbed it and ran with it.

This book began as a talk that I organized around the title "Message from Mother," that I gave at the Gather the Women Congress in Dallas in October 2004. "Mother" worked on me. It could be the third in the series: first there was *The Millionth Circle* then came *Crones Don't Whine* with "Crones Together Can Change the World," and "Exceptional Men Can Be Crones" as chapters. I sent my editor an outline of the idea in December, while in the last stages of an extensive house remodeling and with the holidays approaching. Egad! I would have two months to get this message marked "URGENT" written, to be included in the Fall 2005 catalog. It got done because the creative surf was up, and I was up and out there every morning, catching the wave, and writing it in.

I suggest to my readers that when an idea or an opportunity presents itself, the question, "Is this my assignment?" can only be answered by you. Others may want to involve you in their project, or insist that you manifest an idea that you have. But, something that comes along with your name on it is one that you recognize yourself. *Urgent Message from Mother* will touch a chord if something is already stirring, as I imagine it may be, because when the morphic field is active, there is movement from the collective unconscious into consciousness taking place in receptive individuals.

I think that women who were part of the women's movement and now are of crone age, with a mind and a heart not fully employed, may be realizing that they are waiting for an assignment. Idealistic, altruistic, passionate young women are also feeling and responding to the call to activism. For women of any age, an assignment presents itself as an invitation. Your assignment is what you feel it is. It may be an opportunity that you realize is meant for you. It may come to you as an inspiration or urge. The women's movement that changed the world for American women and rippled out to influence the world was the sum of individual women doing whatever they were moved to do, once they became aware of inequality and injustice. The vessel that supported them through their changes and those they brought into the world was their women's group.

I believe that the thought that women together can change the world is emerging into the minds and hearts of many of us, and that the vessel for personal and planetary evolution is the circle with a spiritual center.

If you have a strong impulse that you can't explain, it could be that poet Olga Broumas is speaking for you in her poem "Artemis" from *Beginning with O.* She wrote:

> I am a woman
> who understands
> the necessity of an impulse whose goal or origin
> still lie beyond me.

At the end of this poem, she says:

> I am a woman committed to
> a politics
> of transliteration, the methodology
> of a mind
> stunned at the suddenly
> possible shifts of meaning—for which
> like amnesiacs
> in a ward on fire, we must
> find words
> or burn.

Women and Goddess became co-opted and lost in the politics of patriarchy; we forgot who we are, and we are now finding pieces, hidden in myths, dug up in archeological sites, uncovered in the Gnostic Gospels. Old Testament meanings shift: "false gods" become images of a divine feminine. The promised land? A land long settled by goddess-worshiping, art-creating, peaceful people who had sacred groves. History shifts: Athens, the cradle of democracy? The cradle of oppressive patriarchy for women. Truth got lost in the transliteration. Women are like amnesiacs who are recovering memory. With nuclear war and overpopulation on the horizon, it is urgent that women awake to the idea that we are the antidote and have the power to change the course patriarchy has set us on. "Gather the women, save the world" is not a tall order when the idea itself may be already moving in the direction of critical mass.

This is Threshold Time

I began writing this book shortly after George W. Bush won his second term of office as president of the United States with the Republican party in control of the Senate and House of Representatives, and the likelihood that during this term he would appoint new Supreme Court justices. The agenda of the Religious Right and Bush's position as a war president received electoral support. Prior to his re-election, bipartisan opposition to funding the development of new tactical nuclear weapons had been thwarted, but whether Congress will now continue to do so is in question. During his first term of office, President Bush withdrew funding from all international programs that included family planning services and supported drilling for oil in the Arctic and logging on public lands, and he considered his narrow political victory as a mandate or political capital he would draw on. These programs support destruction of the environment, population growth, and the proliferation of nuclear weapons.

If only those in power would consider the effects of actions taken now on seven generations to come, or even on the next two generations, as wise clan grandmothers would do!

The United States is in the driver's seat, accelerating toward a bad end for the Earth and its inhabitants, but as long as it is possible to alter the course, there is hope. *Once nuclear bombs and missiles explode and nuclear winter results, or a population explosion exceeds the limits that the Earth can support, there will be no turning back. We aren't there yet.*

This is a crucial time in human history and the life of the planet; a time when tilting the outcome in one direction or the other is possible. Each person doing her or his part in response to something stirring in the morphic field can be that difference. I think of this time as *liminal*, a word derived from the Latin word for threshold. In human affairs, choice matters. Who we become, depends upon the choices we make. At particular times in our lives, really significant choices are made that shape our fate. When we are in a threshold time, what we decide to do determines what comes next. This same principle applies to humanity as the sum of choices made for us or by us.

It is interesting to feel this way and then to learn that this is also a liminal or cusp time on esoteric calendars. Indigenous people in South and Central America whose beliefs are influenced by the Mayan calendar look to 2012 as the end of a 26,000 year cycle. According to Carlos Barrios, an anthropologist and historian, Mayan elders anticipate that this will be a time of transformation when peace begins and people live in harmony with Mother Earth, rather than a destructive end. A Hindu cycle may also coincide, and of course, the Age of Aquarius has been anticipated since the mid-twentieth century as an approaching time of equality and peace. We are currently in the astrological transition zone between the Piscean Age and the Aquarian Age. As bad as the road we are traveling down seems to be, stirrings in the grassroots and alignments in the esoteric heavens support the possibility that we actually could be approaching a positive turning point.

History can also be seen retrospectively as a series of eras, with repression and resistance to change by those that hold to old authority and beliefs on the cusp of the new era. The presidency of George W. Bush is at the center of a transition time in his policies. He is re-entrenching patriarchy by consolidating power and if his efforts to impose the religious beliefs of his political base on everyone else succeed, women will be disempowered. Yet he is also contributing to the possibility of a new era by exporting democracy as foreign policy. Beginning in Afghanistan and Iraq, his administration has demanded elections, which gives immediate suffrage to women and men who have never voted before, and advocated or required women's participation in the governance process.

Envisioning a New Era

There is a rise in consciousness and in the numbers of enlightened men and women in the world who realize that humanity is one family, that women and men must participate equally, and that the Earth is home. Human beings imagine a possibility for themselves and then move toward it. The vision of what is possible comes first, for individuals, nations, and the planet. The founders of the United States drew wisdom from the checks and balances of the indigenous North American people of the Iroquois Confederacy, adopting most of their principles (with the exception of equality of women and the council of grandmothers), when they drafted the American

Constitution, followed by the Bill of Rights as the vision for the new nation.

The image of Earth from outer space is the vision for our time. How to inhabit a beautiful Earth in peace is the alternative to the destructive course of patriarchy. Resolutions drafted and adopted by the United Nations present us with a map.

1. The United Nations Convention on the Elimination of All Forms of Discrimination Against Women (CEDAW) adopted in 1979 by the General Assembly, and signed by 165 nations, with the exception of the United States, North Korea, Afghanistan, and Iran, is often described as a bill of rights for women.
2. The Programme of Action adopted at the International Conference on Population and Development in Cairo in 1994 by acclamation addressed the need for universal education for girls and women, the reduction of infant, child, and maternal mortality, and access to reproductive and sexual health services, including family planning (which is why the Bush administration will not fund UNFPA).
3. The Platform for Action adopted at the UN World Conference on Women (1995) lays out plans to achieve equality and safety for girls and women, and maintains for the first time that women's rights are human rights.
4. The UN Millennium Declaration of Goals (MDG), a global consensus reached by 189 UN member states in 2000, set out to reach eight specific goals in areas of development

and poverty eradication, peace and security, environmental protection, and human rights and democracy.

5. The UN Security Council Resolution 1325 on Women, Peace, and Security (passed unanimously, October 31, 2000), was the first resolution to address the impact of war on women and to acknowledge that women have contributions to make to conflict resolution and sustainable peace.

Apart from these UN resolutions and agreements, there is the Earth Charter. This is an international people's agreement (neither governments, the United Nations, nor organizations) for a compassionate, just, and sustainable world that was written by thousands of people in seventy-eight countries over a twelve-year time period (*www.earthcharter.org*) and was launched at the Hague Peace Palace in June 2000. Its core values are: interdependence, economic and social justice, peace, democracy, and ecological integrity.

In the United States, legislation to create a cabinet level Department of Peace was introduced into Congress in 2004 by Representative Dennis Kucinich and fifty-three cosponsors. The Department of Peace would provide policies, interventions, and programs for domestic and international use, and create a U. S. Peace Academy. This bill will be reintroduced each session and is expected to take years of persistent citizen lobbying to overcome resistance. The long and successful effort to gain women the right to vote inspires its advocates to be persistent and to not be disheartened.

Nuclear Winter

Unless common sense and wisdom rise in time to change the course, those men who can lead us down the path to nuclear catastrophe and overpopulation will do just that. Nuclear winter, described by Carl Sagan as the inevitable consequence of using nuclear weapons, was a deterrent during the Cold War, when the United States and Russia were in a nuclear arms race. When the nonproliferation agreements were signed, the threat seemed over. The Berlin Wall came down soon after that, and the Cold War ended. It turned out to be a remission, not a cure. The threat that nuclear winter poses has also been greatly diluted in our minds, by including nuclear weapons as just one of several types of weapons of mass destruction (WMDs).

The nuclear winter scenario: nuclear bombs and missiles are used; mushroom clouds rise above the target cities; radioactive ash, dust, and soot particles from the destruction are carried on the winds; and the atmosphere becomes thick. A grayish-brown pall covers the entire Earth, and there are no bright sunlit skies, no beautiful sunrises or sunsets, no stars seen from Earth. Nights are black, days are dim. The temperature on the Earth drops, every living thing that depends upon the sun becomes pale and stunted or dies, and anything that survives will experience unremitting nuclear winter. The abundance of life and the beauty of the Earth is gone, the planet becomes a wasteland. If seen from outer space, the planet is no longer a beautiful blue and white sphere. The planet Earth will survive the cataclysmic effects of a nuclear holocaust, and some life forms will probably emerge that can live with radioactivity in the dim light. Cockroaches will probably survive.

Overpopulation

Earth currently has a population of over 6 billion people, doubled since 1960, and is adding about 78 million people every year, or the equivalent of the population of the city of San Francisco every three days. Overpopulation is the biggest ecological threat to the planet. Earth is a garden island in the midst of a vast expanse of space. There is nowhere to go once we use up what sustains us. If we pollute the water, cut down the trees and the rain forests, destroy the ozone layer, turn fertile land into deserts, and continue to create larger, more sprawling, more numerous and unmanageable cities, then Jared Diamond's description of what happened to the people who once inhabited Easter Island will be the model of what could happen to the Earth.

In *Collapse: How Societies Choose to Fail or Succeed,* Diamond describes how deforestation and wind led to a disastrous erosion of the topsoil. The point was reached when there were more mouths to feed than food. Widespread starvation was followed by a descent into cannibalism, and the population that was estimated to be between 6,000 and 30,000 died off. When explorers came upon the almost deserted and barren island, a few survivors and the mute, mysterious, and monumental stone heads facing the sea were all that remained. There had apparently been conflict and competition between priests or chiefs to erect larger and larger stone heads. Civil war and the toppling of the heads preceded the end. The whole forest was gone, and all of Easter Island's tree species were extinct. Among them was the largest palm tree that ever existed. This fact reminds me of California's old-growth redwood trees, which are the tallest living species on Earth. Some are over 2,000 years old and have grown to heights of 300–350 feet. Ninety-five percent of them are already gone.

The Easter Island story is a metaphor for humanity's relationship to Earth. It is a story with a moral to it. Like Easter Islanders, we have no other island/planet to which we can turn for help.

While Diamond does not say how to avert the fate of Easter Island, there is a grassroots solution to the looming problem of overpopulation: Educate women and provide information and access to contraception. It turns out that what is best for the individual woman and her children will collectively be best for the planet. The evidence is in: When women are educated and have reproductive choice, they delay their first pregnancies and have fewer children, who are healthier. A residual from when women were property is the idea that men have dominion over a woman's womb. The rhetoric focuses on abortion, which no women with maternal instinct would ever do lightly. Reproductive rights in general, planning when and how many children to have, and contraception are the real issues. The right to make a decision that will forever change a woman's life is an essential freedom for her and a decision that determines whether a child is wanted and enters the world loved or resented. Marge Piercy in her poem "Right to Life," from *The Moon Is Always Female,* addresses the cost to the child and to the world of forcing women to bear unwanted children:

> We are all born of woman, in the rose
> of the womb we suckled our mother's blood
> and every baby born has a right to love
> like a seedling to sun. Every baby born
> unloved, unwanted is a bill that will come
> due in twenty years with interest, an anger
> that must find a target, a pain that will
> beget pain. A decade downstream a child
> screams, a woman falls, a synagogue is torched,
> a firing squad is summoned, a button
> is pushed and the world burns.

Anything We Love Can Be Saved

I've borrowed this phrase from *Anything We Love Can Be Saved: A Writer's Activism* by Alice Walker. Fierce compassion to protect those who are vulnerable, love of Earth, and outrage at injustice have inspired her words and her actions. There is an activism that I see in Alice, feel in myself, and know to be present in women that is similar to the motivation of a mother or a mother bear.

"Anything we love can be saved" begins with love for what is endangered or is being harmed, hurt, or neglected. Mother love for a child is one model for committed activism. A mother will fight city hall in order to get a disabled child's needs taken care of by an indifferent bureaucracy. Julia Butterfly Hill lived 180 feet up in an ancient redwood tree that she named Luna and loved in order to save an ancient old-growth forest from being logged. She stayed there for 738 days and brought worldwide attention to the plight of these ancient forests. She had no idea when she volunteered to tree sit, which others were also doing, that she would not come down for two years, become a spokesperson, the author of *The Legacy of Luna,* and then the founder of the Circle of Life organization.

Instead of a usual honeymoon, Zainab Salbi and her husband went to Croatia under the sponsorship of a Unitarian church to distribute supplies and money to refugees in the early 1990s. Moved by the plight of women refugees, an intuitive sense of what was needed, and a passion to help, Zainab founded and became president of Women for Women International. This organization has helped women in refugee camps in Afghanistan, Kosovo, Bosnia, Croatia, Colombia, Iraq, Nigeria, Pakistan, Bangladesh, the Democratic Republic of Congo, and Rwanda. The paid staff at the refugee camps are women who, like those they now help, arrived with nothing, separated by war and deaths from home and family, traumatized, most likely raped and ashamed. When it was obvious that the program was very effective (first stage, the women are given money to spend however they wish; second stage, they learn a skill that will enable them to support themselves; third stage is spent in a circle of peers, where stories are a part), appeals to take the program to yet another refugee camp would arrive. Zainab, who was twenty-three when she began and is still young, is the heart of an organization that is on call to help women refugees whenever there is yet another fratricidal war.

When the Bush Adminstration withheld $34 million voted by Congress in July 2002 for the United Nations Population Fund, Lois Abraham, a lawyer in New Mexico and Jane Roberts, a college French teacher in California heard the news and independently reacted. Deciding that it wasn't enough to write a letter to the editor or to a congressman, the two began a grassroots organization 34 Million Friends of UNPFA to make up this funding loss.

For Alice, Julia, Zainab, Lois and Jane, most women activists, and practically all mothers, doing something leads to doing more when your heart is committed. What it takes and how we grow through doing whatever is required for a child or for a cause is maturation ground.

There are many assignments. Some may present themselves and require an immediate, spontaneous response. It may feel like an irrational impulse to say "yes!" in a blink, and yet you had set out in this direction and were in place. Other assignments are for you to pick up or even choose among.

There are many styles and no right way for everyone, but heart and soul need to be engaged for the choice of path to have personal meaning, and most women travel best in the company of others. Each of us has only our allotted time to use as we choose, midlife comes quickly to let us know how fast the years will go by before our stint here is over. Time is speeding up for the planet as well. A worst-case scenario could happen in our lifetime or be what we leave as a fate for our children to face in theirs. The outer world numbs us to what soul has to say, unless we pay attention.

Poems, Prayers, Dreams

Any woman who begins writing poems as a girl and keeps them as one would a private journal stays in touch with what she feels, draws from metaphors, finds a perspective that is deeply personal, and yet can soar above the personal into collective experience. It is a creative and spiritual practice to write poetry. Often women who wrote poetry in adolescence find the urge to do so again later in life, especially during difficult times of transition or when there is an upwelling of deep feelings. At whatever age, under whatever circumstance, if a poem arises in your mind or a desire to write poetry comes to you, treasure the poem or the urge. You honor yourself by writing it down and transferring it to a book with blank pages that you start for this purpose.

I also think that prayers for the well-being of others and yourself bring more love into the visible world from the invisible world, out of which synchronistic events arise. The essential characteristic of love and intercessory prayer is that the more you give away, the more you have, and the more there is. It is the opposite of power and material possessions, where if you give them away, someone else has more and you have less. I suggest that from time to time you focus on what is important and compose a prayer, and then copy it in that same special journal.

Also, take notes on your dreams and take time to write out the big dreams that come to you. These dreams are pregnant with meaning and feeling, may not be readily interpretable, and yet have a power that works on us. Write them down, think about them, return to them later, and tell others with whom you can share your inner life.

These acts feed the soul and inform us. Nourishing the soul is essential to staying on your course, as is finding meaning in what you do.

Stories as Soul Nourishment

The stories that stay with us are another source of soul nourishment. When we take a story to heart, we are inspired by what others have done. Stories about others speak to us about what is possible for us to do as well. They stretch our imagination and therefore the field of what is possible for us to do, grows. Each person's story that I have retold in this book is a teaching story, a part of our lineage, and yet they are mostly about our contemporaries. They are women who are in our invisible circle of sisters, women whose protective maternal instinct, feminine wisdom, or outrage motivated them to act. They are also archetypal figures in that their qualities live within us.

Stories that have vitality in them, that go back hundreds or even thousands of years, are to the culture what big dreams are to the individual. They have levels of meaning and can be returned to again and again, as another aspect becomes significant, another layer of interpretation makes sense. One of the most important to me is the Grail legend.

Return of the Grail to the World

The Grail story which Chretien de Troyes started to write (and never finished) in the twelfth century was the simplest and most symbolically relevant to our times. The cast includes Perceval, the knight who doesn't ask questions; a king with a wound that will not heal; a woman who carries the Grail; and the mysterious, numinous Grail itself. The three come together one evening in the Grail castle.

Perceval was returning to the forest to find his mother, when he comes to a place in the river, too deep for him to cross. There he saw a man fishing in a boat, who invites him to spend the night. Perceval becomes a guest in the castle of the Fisher King and sits down to a multi-course dinner in the large hall.

Between each course, a procession enters the hall and passes through. The last figure in every procession is the Grail carrier. When she enters with the Grail, it is clearly something wondrous and awesome. Perceval, of course, notices all this. But he remembers that his mentor told him, "Don't ask questions," thinks to himself that he will ask one of the servants later, and makes small talk over dinner. The next morning, he awakens to an empty castle and once he crosses over the drawbridge, the castle itself disappears.

Perceval learns too late what this was about and how he failed. The king has a wound that will not heal and a kingdom that is a wasteland. Only when his wound is healed will his kingdom be restored, and only the Grail can heal him. Perceval saw the wounded king and saw the Grail, but failed to ask "What ails thee?" of the king; a question that a compassionate person might ask and one that would lead to learning the cause or nature of the problem. When he was in the presence of the Grail, he failed to ask "Whom does the Grail serve?" and so he did not learn that it was a sacred vessel that could restore wholeness and heal.

The Grail is usually thought of as a chalice, which is a feminine symbol. A feminine symbol that is numinous, mysterious, awesome, wondrous and healing, that is hidden in the forest/the collective unconscious and carried by a woman is a symbol of the Sacred Feminine. Under patriarchy, the rites of the Goddess and the rights of women disappeared together. Both were suppressed and forgotten.

I first described the Grail as a Christ symbol and as a symbol of the Self in *The Tao of Psychology: Synchronicity and the Self.* Then I went on a pilgrimage that was both an outer journey and an inner one. I thought about the difference between the transcendent divinity of God, which is the only aspect of divinity recognized by patriarchal religions, and embodied spirituality, which is of the Goddess, felt in the body in Nature and through sacred and ecstatic experiences that can occur in childbirth, nursing an infant, making love, dancing, and at sacred sites. This led me to think again of the Grail legend and write *Crossing to Avalon: A Woman's Midlife Quest for the Sacred Feminine.*

The Grail is most often thought of as the cup filled with wine that was used by Jesus at the Last Supper when he said "This is my blood . . . do this in remembrance of me," which is the basis of the Christian sacrament of communion. While it is a symbol associated with Jesus, it is archetypally a feminine symbol, and when filled with sacred blood, is easily understood as a symbol of the Goddess. Both are aspects of the Self, which is called God, Goddess, Tao, Higher Power, the Great Mystery, and by the myriad names that humanity has had for divinity, and prior to the imposition of patriarchy was not exclusively male.

The Sacred Feminine is returning to the world through ordinary women who are carrying the healing power of the feminine and the Goddess back into the world. They are the Grail carriers. Ordinary women are also contemporary Percevals who are asking the right questions and seeing the wounded king whose kingdom is a wasteland as a symbol of patriarchy.

Intuitions and Seeds

Intuitive insight comes in many ways, including an inner voice that comes through loud and clear to let you know when an assignment is yours.

Your heart may do the selecting. You may find yourself unable to walk away from someone who needs something that you want to provide, or away from something you want to do. And in either case, you don't even know if you can. When your heart is your guide, this assignment has your name on it.

Like an unexpected pregnancy, you may be carrying the seeds of something that will change your life and the world. Take that seed and bring it into a circle of women, nurture it with wisdom, give it energy, prune what needs pruning, let the tap root go down into the energy field of Mother Earth, to draw from and contribute thought and action to the morphic field, and then take it out into the world to bloom and bear fruit.

Every single thing in Nature belongs to its particular group, with which it shares similarities and yet is unique, with no two of anything alike. Yet each comes into bloom or fruition with the others, in season. Some species can stay dormant for long stretches of time, waiting for just the right conditions, and then all at once seeds unconnected to each other directly begin to send tendrils up to the surface. To the unobservant, when they do bloom, it is as if they appeared overnight.

I think this is analogous to what is happening now. Attention goes to where the action is—on wars and conflicts, on centers of power, on scandals and celebrities. Unnoticed and still very close to the ground, a message is rising into consciousness. It is growing more in some places than others. Invisibly linked like communication on the Internet, or like bilocality noted in physics where related particles separated by vast distances move together, or like wellsprings drawing from the same aquifer, women are getting the message:

Gather the women, save the world.

ACKNOWLEDGMENTS

The conveners of the Millionth Circle are the godmothers of this book:

Katherine Collis, Elinore Detiger, Donna Goodman, Nancy Grandfield, Judy Grosch, Ronita Johnson, Betty Karr, Joan Kenley. Leslie Lanes, Penny McManigal, Linda Merryman, Clare Peterson, Elly Pradervand, Betty Rothenberger, Peggy Sebera, Ann Smith, Onnolee Stevens, Justine Toms, Andrea Wachter, Joan Whitacre, Ingrid Willgren, and Rosemary Williams.

Dennis Kucinich drew my attention to the original Mother's Day Proclamation. Patricia Smith Melton took the photographs of the women on the book jacket while fostering international connections between women's circles. "Gather Women," the words that initially reverberated in my psyche, came to me through Carol Hansen Grey and Marilyn Nyborg. The invitation to speak at the Gather the Women Congress organized by Donna Collins and Kathe Schaaf, led directly to writing this book.

The archetypal psychology of C.G. Jung continues to be the foundation for my understanding of the individual and the collective psyche; theoretical biologist Rupert Sheldrakes's idea of morphogenesis added a dimension to Jung's psychology, applicable to activism. The study of the Old and New Testament at Pomona College broadened my knowledge and interest in when and how scripture became scripture as well as reading these texts through. Neurology and the endocrinology that were part of my medical curriculum and psychiatric training proved useful in scanning new research that points to differences between men and women.

As the ideas in *Urgent Message From Mother* came into my mind and onto the page, I thought of books and people who influenced me by their words or example: Angeles Arrien, Helen Caldicott, the Dalai Lama, Riane Eisler, Clarissa Pinkola Estés, Betty Friedan, China Galland, Susan Griffin, Carol Gilligan, Marija Gimbutas, Nelson Mandela, Elaine Pagels, Charlene Spretnak, Merlin Stone, Deborah Tannen, Gloria Steinem, Desmond Tutu, Alice Walker, Marianne Williamson, and Marie Wilson. Anne Dosher, May East, Beverly Engel, Glenna Gerard, Judy Grosch, Stephanie Hiller, and Angela Weber helped to further the Millionth Circle vision. For miscellaneous good reasons, I also want to thank Valerie Andrews, Barbara Hort, Jananne Lovett-Keen, Sharon Medhi, Louise Vance, Jo Wharton, and my agent, Katinka Matson, president of Brockman, Inc.

It is a pleasure to work with "Team Red Wheel", my affectionate name for Jan Johnson—my editor, Michael Kerber, Brenda Knight, Jill Rogers, Bonni Hamilton, Kathleen Wilson Fivel, Emily Logan, Rachel Leach, Kate Hartke, Stephanie Tagg, and Liz Wood.

I feel a deep sense of gratitude for my home base circle with a spiritual center, my sisters: Isabel Allende, Grace Damman, Carole Robinson, Pauline Tesler, and Toni Triest.

Lastly, with time of the essence, I was greatly helped by grace and by my research assistant, Google.

DISCUSSION GUIDE AND CIRCLE EXPERIENCE

Dear Book Friends,

If your book club decides to read and discuss *Urgent Message from Mother*, which encourages the formation of circles with a spiritual center, I encourage you to have a book discussion *and* a circle experience. With imagination and intention, the circle experience can take only a couple of minutes at the beginning and end of the club meeting. I suggest the following. Imagine that you are meeting around a fire, or light a candle in the center of your group as a symbol of warmth and illumination. Have a moment or two of silence before starting the discussion. In the silence, each woman might take a couple of deep breaths to feel a center in herself where wisdom and receptivity dwell. Let a word or phrase come to mind to describe how each of you is feeling in this moment. Then do a short and simple check-in. Go around the circle: each woman in turn says her word or phrase. Then begin the discussion as usual. Or consider using circle principles (download from *www.millionthcircle.org* or gleaned from *The Millionth Circle*). At the end of the meeting, take another moment for silence, let another word or phrase come to mind, and check out by saying this word or phrase into the circle. When the last woman speaks, the circle is ended by blowing out the candle, or saying something simple, or if the circle decides to hold hands at the end, with a squeeze that goes hand to hand around the circle.

The following are suggestions for discussion.

Introduction

Do the words "Gather the women" call to you in some way? Compare an aspect of life at the beginning of the twenty-first century with the world that your mother or grandmother lived in, when they were the age you are now.

Chapter 1 Mother's Day

Do you agree that there are gender differences in how women and men react to war? If grandmothers decided whether a country went to war, what would you expect might happen? Do you see a relationship between traumatized children and terrorist behavior?

Chapter 2 Mother Earth/Mother Goddess

Is the premise that before there was God, there was Goddess new to you? Speculate on how conceptualizing divinity as Mother instead of Father would make a difference. Does the Gaia Hypothesis or the image of Earth from outer space cause you to relate to the Earth or the state of the world differently than you once did?

Chapter 3 Monotheism/Doing without Mother

Can you see a relationship between dysfunctional family patterns and fratricidal wars, as the author maintains there is? What are your thoughts about the familiar stories of Abraham and Isaac and Cain and Abel when seen through a psychological lens? In your experience, is there a difference between religion and spirituality? Is there a gulf between the political positions taken by leaders of your religion and the compassionate meaning that your religion has for you?

Chapter 4 The Problem with Patriarchy

After learning about women's "tend and befriend" response to stress, can you recall examples when you responded in this way? If men use conversation to determine who is alpha to whom and women use conversation to bond, what happens in conversations between men and women? Have you seen examples of bullying, and do you have some insight into this? Does your own experience coincide with the gender differences that the author suggests? Politically incorrect? Female chauvinism?

Chapter 5 Antidote 1:
The Visible Power of Women Together

Do you think that an antidote to patriarchy is needed? Did a particular example that the author presented about the power of women together remind you of a personal story? Did a particular story inspire or suggest something you might want to do? What might you do if you were brave?

Chapter 6 Antidote 2:
The Invisible Power of Women's Circles

Discuss the idea of critical mass: the metaphoric millionth circle, the tipping point, and morphic field theory. Is it plausible that forming a circle has invisible effects?

Chapter 7 Anything We Love Can Be Saved

Do you agree that women as a gender possess qualities that are needed to make the world safe? Do you feel that the nuclear winter or Easter Island scenarios are likely risks, or the words of an alarmist? Looking back on your life, have there been times when you felt that you were fulfilling your "assignment"? Do you have one now?

CHAPTER RESOURCES

Introduction

Circles of Compassion and November 2, World Day of Circles of Compassion: Women's World Summit Foundation, *www.woman.ch.*

Gather the Women: *www.gatherthewomen.org.*

Millionth Circle: *www.millionthcircle.org.*

PeaceXPeace ("peace by peace"): *www.peacexpeace.org.*

Women's World Summit Foundation: *www.woman.ch.*

Chapter 1 Mother's Day

Amma (Sri Mata Amritanandamayi Devi). *The Awakening of Universal Motherhood.* Kerala, India: Mata Amritanandamayi Mission Trust, 2003.

Amnesty International. *It's in Our Hands: Stop Violence against Women.* 2005.

Baron-Cohen, Simon. *The Essential Difference: Men, Women and the Extreme Male Brain.* New York: Penguin, 2003. (Questions from "Girls, Boys and Autism." *Newsweek,* September 9, 2003.)

Beijing Declaration and Platform for Action.

Code Pink: *www.codepink4peace.org.*

Ending Violence against Women. Baltimore: Johns Hopkins School of Public Health, Population Reports No. 11, 1999.

Fagan, Kevin. "A New Battle of the Sexes: Men and Women View the War and Its Coverage with Fundamental Differences." *San Francisco Chronicle,* March 28, 2003.

Fetal and Neonatal Stress Research Group, Faculty of Medicine, Imperial College School of Medicine, London. Report online, 2004 at *www1.imperial.ac.uk/medicine*. Summary of research with selected publications.
Friedan, Betty. *The Feminine Mystique.* New York: W. W. Norton, 1966.
Global Exchange: *www.globalexchange.org.*
Global Peace Initiative of Women: *www.gpiw.org.*
Million Mom March: *www.millionmommarch.org.*
Mothers Against Drunk Driving: *www.madd.org.*
Mother's Day Proclamation: *www.peace.ca/mothersdayproclamation.htm.*
Narhaniesz, Peter W. *Life in the Womb: The Origin of Health and Disease.* Ithaca, NY: Promethean Press, 1999.
Report from President's Commission on the Status of Women (1961–1963) *www.lexisnexis.com/academic/2upa/Aph/pcmStatusWomen.asp.*
Women's International League for Peace and Freedom: *www.wilpf.org.*

Chapter 2 Mother Earth/Mother Goddess

Allen, L. S., Richey, M. F., Chai, Y. M., Gorski, R. A. "Sex Differences in the Corpus Callosum of the Living Human Being." *Journal of Neurosciences,* April 11, 1991, pp. 933–42.
Allison, Teri Wills, mother of a soldier in Iraq. *San Francisco Chronicle,* November 21, 2004.
Biblical References: The Holy Bible: Revised Standard Version. New York: Thomas Nelson, 1953. First and second commandments: Exodus 20: 2–4, Deuteronomy 5:7–8.
Bolen, J.S. "Aphrodite: Goddess of Love and Beauty, Creative Woman and Lover" in *Goddesses in Everywoman: Powerful Archetypes in Women's Lives.* San Francisco: HarperSanFrancisco, 2004.
Carson, Rachel. *Silent Spring.* Boston: Mariner Books, 2002.
Ciabattari, Jane. "From Rwanda's Ashes, Women Are Building Anew." January 16, 2005, women's enews: *www.womensenews.org.*
Eisler, Riane. *The Chalice and the Blade.* San Francisco: Harper & Row, 1987.

Gimbutas, Marija. *Goddesses and Gods of Old Europe, 7000–3500 BC.* Berkeley: University of California Press, 1982.

Greenpeace: *www.greenpeace.org.*

Lovelock, J. E. *The Ages of Gaia: A Biography of Our Living Earth.* New York: W. W. Norton, 1988.

Millenium Ecosystem Assessment: *www.milleniumecosystemassessment.org.*

Stone, Merlin. *When God Was a Woman.* Irvine, CA: Harvest Books, 1978.

Shlain, Leonard. *The Alphabet Versus the Goddess.* New York: Penguin Putnam, 1998, pp. 82–83.

Walker, Alice. "We Have a Beautiful Mother," from *Anything We Love Can Be Saved.* New York: Ballantine Books, 1997.

Chapter 3 Monotheism/Doing without Mother

Biblical References: *The Holy Bible: Revised Standard Version.* New York: Thomas Nelson, 1953. Creation of male and female together: Genesis 1:27, 5:2; Creation of Eve from Adam's rib: Genesis 2:22; Cain and Abel: Genesis 4:2–9; Reason for the Flood: Genesis 6:6–7; God's promise to not destroy every living creature again: Genesis 8:21, 9:11; Destruction of Sodom and Gomorrah: Genesis 19:24; Abraham's willingness to sacrifice Isaac: Genesis 22: 1–13; Hagar and Ishmael: Genesis 21: 14–21; Do justice, love kindness, walk humbly with your God: Micah 6:8. The two great commandments (love God, love neighbor): Matthew 22: 37–40, Mark 12: 30–31; What you do to the least of them: Matthew 25:40; Judge not: Matthew 7:1, Luke 6:37; Do to others as you would have them do to you: Matthew 7:12; My God, My God, why have you forsaken me? Matthew 27:46; Reconciliation with God through Jesus Christ: Romans 5:9–11.

Ciabattari, Jane. "From Rwanda's Ashes, Women Are Building Anew." January 16, 2005, women's enews: *www.womensenews.org* (Source for effect of Belgian colonial favoritism).

Interfaith Encounter: *www.interfaith-encounter.org.*

Pagels, Elaine. *The Gnostic Gospels.* New York: Random House, 1979.
United Religious Initiative: *www.uri.org.*

Chapter 4 Mother Needs You!

Gilligan, Carol. *In a Different Voice.* Cambridge, MA: Harvard University Press, 1982.

Goldstein, Joshua. *War and Gender: How Gender Shapes the War System and Vice Versa.* Cambridge: Cambridge University Press, 2001.

Limbaugh, Rush: *www.thisistheshit.org* (May 6, 2004, transcript and MP3 of exchange between caller and Limbaugh).

Northern Ireland Women's Coalition: *www.niwc.org.*

Ray, Paul H., and Sherry R. Anderson. *The Cultural Creatives.* New York: Harmony Books, 2000.

Tannen, Deborah. *You Just Don't Understand: Women and Men in Conversation.* New York: Ballantine, 1990.

Taylor, S. E., Klein, L. C., Lewis, B. P., Gurung, R. A. R., Gruenewald, T. L., Updegraff, J. A. "Female Responses to Stress: Tend and Befriend, Not Flight or Fight." *Psychological Review,* 2000, pp. 411–429.

Time. Persons of the Year 2002, December 22, 2002.

Women's Environmental and Development Organization (WEDO). *Beijing Betrayed,* 2005: *www.wedo.org.*

Chapter 5 Antidote 1: The Visible Power of Women Together

Angier, Natalie. "Bonobo Society: Amicable, Amorous, and Run by Females." *New York Times,* April 22, 1997.

Boston College Center on Wealth and Philanthropy (for wealth transfer reports): *www.be.edu/research/swri.*

Ciabattari, Jane. "From Rwanda's Ashes, Women Are Building Anew." January 16, 2005, women's enews: *www.womensenews.org.*

De Waal, Frans. *Bonobo: The Forgotten Ape.* Berkeley: University of California Press, 1997.

Emily's List: *www.emilyslist.org.*

Family Violence Prevention Fund: *www.endabuse.org.*

Feminist Majority Fund: *www.feminist.org.*

Galland, China. *The Bond between Women.* New York: Riverhead, 1998, p. 209.

Ms. Foundation for Women: *www.ms.foundation.org.*

Ms. magazine: *www.msmagazine.com.*

Rosen, Ruth. "The Power of Peaceful Protest." *San Francisco Chronicle,* July 22, 2002.

Stolberg, Sheryl Gay. "9/11 Widows Skillfully Applied the Power of a Question: Why?" *New York Times,* April 1, 2004.

Tjaden, Patricia, and Nancy Thoennes. *Extent, Nature and Consequences of Violence against Women: Findings from the National Violence against Women Survey.* Atlanta: National Institute of Justice and the Centers for Disease Control and Prevention, 2000.

Vote, Run, Lead Initiative: *www.voterunlead.org.*

The White House Project: *www.thewhitehouseproject.org.*

Women/51.3 percent wealth information: *www.pbs.org/ttc/society/philanthropy.html.*

Women's Funding Network: *www.wfnet.org.*

Chapter 6 Antidote 2: The Invisible Power of Women's Circles

Estés, Clarissa Pinkola. "Do Not Lose Heart." *www.creativeresistance.ca/strength/do-not-lose-heart-clarissa-pinkola-estes.htm.*

Gladwell, Malcolm. *The Tipping Point: How Little Things Can Make a Big Difference.* Boston: Little, Brown, 2002, pp. 10, 258, 259.

Kauter, Karl. "A Tale for All Seasons." *www.storybin.com.*

Sheldrake, Rupert. *The Presence of the Past: Morphic Resonance and the Habits of Nature.* Rochester, VT: Park Street Press, 1988, 1995, pp. 177–78.

Chapter 7 Anything We Love Can Be Saved

Barrios, Carlos. Saq'Be' Organization for Mayan and Indigenous Spiritual Studies, *www.sacredroad.org.*

Bolen, J.S. *Crossing to Avalon,* "The Grail Legend: The Spiritual Journey," "Women's Mysteries and the Grail." San Francisco: HarperSanFrancisco, 2004.

Broumas, Olga. "Artemis," from *Beginning with O.* Yale Series of Younger Poets, Vol. 72. New Haven, CT: Yale University Press, 1977.

Department of Peace activism: *www.dopcampaign.org.*

de Troyes, Chrétien. *Perceval or The Story of the Grail.* Translated by Ruth Harwood Cline. Athens, GA: University of Georgia Press, 1985.

Diamond, Jared. *Collapse: How Societies Fail or Succeed.* New York: Viking, 2005.

Piercy, Marge. "Right to Life," from *The Moon Is Always Female.* New York: Alfred A. Knopf, 1977.

Redwood trees: *www.sempervirens.org.*

Sagan, C. *The Nuclear Winter: The World after Nuclear War.* London: Sidgewick and Jackson, 1985.

34 Million Friends of UNPFA: *www.34millionfriends.org* (UNPFA is the United Nations Population Fund).

Walker, Alice. *Anything We Love Can Be Saved: A Writer's Activism.* New York: Random House, 1997.

World population and overpopulation information, see World Population Awareness: *www.overpopulation.org.*

World population figures: *http://www.census.gov/ipc/www/popclockworld.html.*

INDEX

PERMISSIONS

Every effort has been made to contact permissions holders of any copyrighted material in this book. If any required acknowledgments have been omitted, it is unintentional. If notified, the publishers will be pleased to rectify any omission in future editions.

"We Have a Beautiful Mother" poem from *Her Blue Body Everything We Know* by Alice Walker, copyright © 1991 by Alice Walker is used by permission of Harcourt, Inc. and David Higham Associates.

Excerpt from the poem "Artemis" from *Beginning with O* by Olga Broumas, copyright © 1977 by Olga Broumas is used by permission of Yale University Press.